Illustrated Genera of Ascomycetes

Volume I

Richard T. Hanlin

with drawings by
Carol Gubbins Hahn

APS PRESS
The American Phytopathological Society
St. Paul, Minnesota

To my parents

This book has been reproduced directly from computer-generated
copy submitted in final form to APS Press by the author.
No editing or proofreading has been done by the Press.

Library of Congress Catalog Card Number: 89-82674
International Standard Book Number: 0-89054-107-8

Second printing, 1990
Third printing, 1992
Fourth printing, 1997

Printed in the United States of America

The American Phytopathological Society
3340 Pilot Knob Road
St. Paul, Minnesota 55121-2097, USA

PREFACE

The idea for this book was conceived a number of years ago, and in the ensuing years materials for the text have been slowly accumulated. Actual work on the manuscript began in mid-1987.

Any work of this kind necessarily draws upon the efforts of numerous individuals whose works were consulted in compiling the generic descriptions; they are cited in the references and their contributions are gratefully acknowledged and appreciated. At least one species in each of the 100 genera included in this work, however, has been personally examined, and all of the drawings are originals made for this book. Likewise, all of the photographs were taken by the author.

Numerous difficulties confront anyone attempting to prepare a work such as this, not the least of which is the considerable disagreement among specialists on the delimitation of genera and sometimes of the generic names themselves. A greater problem, however, is the lack of adequate descriptions in the literature for most genera of ascomycetes, so that descriptions often had to be compiled from several sources. The lack of data has resulted in a certain amount of uneveness in the descriptions. The selection of the genera to be included and the information contained in the descriptions is the sole responsibility of the author.

Various persons have contributed in different ways to the development of the key, especially students in my classes over the years who have used different versions and made useful comments regarding them. Several individuals, however, deserve special mention for their role in making this book a reality. I am most grateful to botanical artist Carol Gubbins Hahn for preparation of the drawings that comprise the heart of the book; her arrival in Athens at the time the project was initiated was most fortuitous. I also thank J. W. Kimbrough and Kathleen Thomson Cason for reviewing the

final manuscript. Donna Jones did the formatting and printing of the camera-ready copy. The project was made possible through the approval and support of Director Charles W. Laughlin of the Athens Experiment Station of the University of Georgia College of Agriculture, which provided some of the funding for the project. Additional funding was provided by National Science Foundation - U.S.-Venezuela Cooperative Project INT-8501713. And I thank APS Press for their willingness to undertake publication of the book.

This book is gratefully dedicated to my parents, whose patient tolerance of my youthful collecting excursions allowed an early interest in nature to develop into an enjoyable and rewarding career. Their constant support and encouragement is deeply appreciated.

Richard T. Hanlin

Athens, Georgia
June, 1989

TABLE OF CONTENTS

ALLANTOSPORAE

PHAEOSPORAE

HYALODIDYMAE

PHAEODIDYMAE

SCOLECOSPORAE

HYALOPHRAGMIAE

PHAEOPHRAGMIAE

HYALODICTYAE

PHAEODICTYAE

INTRODUCTION

This book arose from a long-observed need for an illustrated key to genera of ascomycetes that are frequently encountered in mycology or plant pathology courses. It is intended not for the ascomycete specialist, but for advanced students of mycology or others who need to identify ascomycetes and who will benefit from having illustrations with which to compare their specimens. The key was originally written as a key to genera of ascomycetes isolated from peanuts; this was later expanded, with the addition of generic descriptions, and computerized for use in mycology classes. A modification of the key was published as a key to genera of plant pathogenic ascomycetes in Venezuela. It is hoped that the addition of illustrations of a representative species for each genus will enhance its usefulness.

A key to 100 genera of ascomycetes is provided, followed by descriptions and illustrations of each genus. Of the 100 genera, 65 include at least one species that is pathogenic to plants, and several other genera are associated with plant materials. An additional 134 genera are discussed under the "Comments" section of each description; these are listed in the index.

The initial couplets in the key are based on Saccardoan spore types, followed by the type of ascoma formed. No cleistothecial plectomycetes or hemiascomycetes are included in the key. Genera are also arranged according to ascospore type, in the following order: hyalosporae, allantosporae, phaeosporae, hyalodidymae, phaeodidymae, scolecosporae, hyalophragmiae, phaeophragmiae, hyalodictyae, and phaeodictyae. Since the key is completely artificial, the arrangement of genera does not imply phylogenetic relationships among them.

The heart of the book consists of the generic descriptions and the accompanying illustrations; these are placed on facing pages for ready reference. The format of the descriptions is as follows: genus and author citation, a detailed generic description, associated genera of anamorphs, habitat, representative species, comments on related genera, and references. The references are not intended to be complete, but rather to serve as a starting point for looking up species descriptions. Emphasis is placed on references that include illustrations of species, and those with keys and descriptions. Full citations for all references are provided at the back of the book.

Because this book is intended for use by students, some comments on use of the key may be helpful. The effective use of any key depends on the user's acquiring an understanding of the terminology and concepts utilized by the author in the construction of the key. This can only be done through experience, but comparison of illustrations of genera with particular characters can be helpful in familiarizing oneself with these concepts (e.g., whether an ascospore is filiform or narrowly phragmosporous). It should be kept in mind that a key is only an aid to identification and that final determinations can be accurately made only after comparing the specimen in question with the generic description. It is at this point that the illustrations should be especially helpful. Uncertainty in choosing one or the other member of a particular couplet suggests that both should be followed; the key is strictly dichotomous. The key is also reversible to permit easy retracing of the steps taken to reach a particular determination. This can also be a useful learning device, as one can start with a known genus and follow the key backwards to see which character choices distinguish that particular genus.

Some problem areas should also be mentioned. Ascospore color is a potentially confusing character. In a number of genera with "hyaline" ascospores, the spores have a greenish tint when observed under the light microscope; such spores must be keyed out as hyaline. (True green is a rare color in fungi). Likewise, ascospores that are very light brown, or "smokey", must be keyed out as brown spores; the color will usually be more apparent in masses of spores. Also, in a number of genera with distinctly hyaline ascospores, it is not uncommon for the spores to acquire a brownish color after discharge. Thus the color of the "mature" spore is the color of the spore in the ascus before discharge. In some genera which characteristically have unicellular spores a septum will occasionally form, especially prior to spore germination. And in some genera with phragmosporous spores, longitudinal septa may occur randomly. Observations of a number of spores will show which is the dominant pattern for a given specimen.

One of the greatest difficulties in constructing a key to genera is to allow for variations among species within given genera. This problem is compounded by the lack of monographs or even species lists for most genera, making it difficult to acquire an understanding of species variation. In *Balansia*, e.g., most species form a flat perithecial stroma on the host, yet a few tropical species form stipitate-capitate stromata. Such diversity must be incorporated into the key if both types of stromata are to be correctly identified. Where such diversity is known, it has been included in the present key. A similar caution must be given for the illustrations. Since only one, or occasionally two, species is illustrated for each genus, not all variations within the genus can be given. Careful comparison of material with the generic description should resolve such problems.

Although the identification of ascomycetes is rarely easy, it is hoped that this book will be a useful aid to anyone who has an interest in, or a need to identify common genera.

KEY TO GENERA

1 Ascospores filiform 77
1' Ascospores other than filiform in shape 2

2(1') Ascospores allantoid 35
2'(1') Ascospores other than allantoid in shape . . . 3

3(2') Ascospores 1-celled 4
3'(2') Ascospores with more than one cell 58

4(3) Ascospores hyaline or greenish 5
4'(3) Ascospores light to dark brown 39

5(4) Ascoma perithecioid or cleistothecioid 6
5'(4) Ascoma apothecioid 20

6(5) Ascoma lacking an ostiole 7
6'(5) Ascoma ostiolate 14

7(6) Ascomal wall thin and hyaline . . . *Brasiliomyces (22)**
7'(6) Ascomal wall thick and dark brown 8

8(7') Ascoma containing a single ascus 9
8'(7') Ascoma containing several-many asci 10

9(8) Ascomal appendages
mycelioid . . . . *Sphaerotheca (32)*
9'(8) Ascomal appendages
dichotomously branched . . . *Podosphaera (30)*

10(8') Setose mycelial cells
formed around ascomata . . . . *Blumeria (24)*
10'(8') Setose mycelial cells lacking 11

11(10') Ascomal appendage mycelioid or
lanceolate 12
11'(10') Ascomal appendage tips dichotomously

branched or curved 13

12(11) Ascomal appendages
mycelioid . . . . *Erysiphe (26)*
12'(11) Ascomal appendages rigid,
tapered, with bulbous
base (lanceolate) . . . . *Phyllactinia (34)*

13(11') Ascomal appendage tips
dichotomously branched . . *Microsphaera (28)*
13'(11') Ascomal appendage tips
curved (uncinate) . . . . *Uncinula (36)*

14(6') Ascus wall evanescent . . . . . . *Ceratocystis (18)*
14'(6') Ascus wall persistent 15

15(14') Ascus unitunicate 16
15'(14') Ascus bitunicate 19

16(15) Clypeus present around
ostiolar neck 17
16'(15) Clypeus lacking 18

17(16) Clypeus poorly developed . . . . . . *Glomerella (40)*
17'(16) Clypeus well developed . . . . *Phyllachora (42)*

18(16') Ascus widest in middle . . . *Physalospora (44)*
18'(16') Ascus clavate or
cylindrical . . . . *Glomerella (40)*

19(15') Ascostroma uniloculate . . . . *Guignardia (46)*
19'(15') Ascostroma multiloculate . . . . *Botryosphaeria (48)*

20(5') Apothecium stipitate, and pileus
capitate, saddle-shaped, or
sponge-like 21
20'(5') Apothecium sessile, or if stipitate,
then cupulate 23

21(20) Ascomal pileus sponge-like . . . . *Morchella (66)*
21'(20) Ascomal pileus capitate or saddle-shaped 22

22(21') Pileus capitate . . . . *Leotia (68)*
22'(21') Pileus saddle-shaped . . . . *Helvella (70)*

23(20') Ascus operculate 24
23'(20') Ascus inoperculate 28

24(23) Ascus apex blueing in iodine . . . . . *Peziza (52)*
24'(23) Ascus apex not blueing in iodine 25

25(24') Apothecium red, orange or yellow 26
25'(24') Apothecium buff, brown or black 27

26(25) Ascospore wall reticulate or
warted . . . . *Aleuria (50)*
26'(25) Ascospore wall smooth . . . . *Sarcoscypha (54)*

27(25') Apothecium light brown, cupulate,
stipe ribbed . . . . *Helvella* (70)
27'(25') Apothecium dark brown to black,
stipe smooth lacking ribs . . . . *Urnula (56)*

28(23') Ascoma stipitate 29
28'(23') Ascoma sessile embedded, or
erumpent from host 32

29(28) Apothecium arising from a sclerotium or
stromatized host tissue, brown 30
29'(28) Apothecia not associated with sclerotia
or stromatized tissues, light
colored . . . . *Hymenoscyphus (58)*

30(28) Apothecia arising from a
sclerotium . . . . *Sclerotinia (64)*
30'(28) Apothecia arising from stromatized
host tissues 31

31(30') Apothecia on mummified fruits .... *Monilinia (62)*
31'(30') Apothecia on flowers, catkins, or stromatized ovaries .... *Ciboria (60)*

32(28') Ascoma elliptical .... *Ploioderma (76)*
32'(28') Ascoma circular 33

33(32') Ascoma circular, formed in a stroma .... *Phacidium (78)*
33'(32') Ascoma circular, not stromatic 34

34(33') Excipulum lacking at sides of apothecium ... *Pseudopeziza (74)*
34'(33') Excipulum well developed at sides of apothecium .... *Blumeriella (72)*

35(2) Stroma valsoid (necks converge) 36
35'(2) Stroma diatrypoid or eutypoid (necks straight) 37

36(35) Blackened zone outlining stroma in host .... *Eutypella (84)*
36'(35) Blackened zone lacking .... *Valsa (86)*

37(35') Stroma eutypoid .... *Eutypa (82)*
37'(35') Stroma diatrypoid 38

38(37') Stroma bright-colored .... *Endothia (88)*
38'(37') Stroma black .... *Diatrype (80)*

39(4') Ascus bitunicate .... *Botryosphaeria (48)*
39'(4') Ascus unitunicate or prototunicate 40

40(39') Ascoma without an opening 41
40'(39') Ascoma perithecioid or apothecioid 42

41(40) Ascus wall persistent .... *Tuber (126)*

41'(40) Ascus wall evanescent . . . . *Thielavia (104)*

42(40') Ascoma an apothecium . . . . *Ascobolus (124)*
42'(40') Ascoma a perithecium 43

43(42') Ascospores truncate at one end 71
43'(42') Ascospores the same at both ends 44

44(43') Ascus wall evanescent 45
44'(43') Ascus wall persistent 49

45(44) Ascospores with a germ slit . . . . *Ascotricha (102)*
45'(44) Ascospores with germ pores 46

46(45') Ascospores with one germ pore 47
46'(45') Ascospores with two germ pores 48

47(46) Ascospores yellowish to straw-colored,
dextrinoid when young . . . . . *Microascus (90)*
47'(46) Ascospores olive-brown to dark
brown, not dextrinoid . . . . . *Chaetomium (100)*

48(46') Ascospores yellowish to
light brown . . . . *Lophotrichus (98)*
48'(46') Ascospores dark brown . . . *Melanospora (92)*

49(44') Perithecia in or on a stroma 55
49'(44') Perithecia not associated
with a stroma 50

50(49') Ascoma red to orange . *Neocosmospora (96)*
50'(49') Ascoma brown to black 51

51(50') Ascospores with germ pores 52
51'(50') Ascospores with a germ slit 54

52(51) Ascospores with a gelatinous
sheath . . . . *Sordaria (106)*
52'(51) Ascospores lacking a sheath 53

53(52') Ascospore wall striate . . . . *Neurospora (108)*
53'(52') Ascospore wall pitted
or reticulate . . . . *Gelasinospora (110)*

54(51') Ascoma usually seated on a
distinct subiculum, perithecia
usually >0.5 mm diam, ascospores
not flattened . . . . *Rosellinia (114)*
54'(51') Ascoma not on a subiculum,
perithecia usually <0.05 mm diam,
ascospores flattened . . *Coniochaeta (112)*

55(49) Stroma erect 56
55'(49) Stroma effuse, pulvinate or
hemisphaerical 57

56(55) Perithecia in a cluster or disc
at apex of stroma,
usually on dung . . . . . . *Podosordaria (116)*
56'(55) Perithecia distributed over
entire surface of stroma
usually on wood or leaves . . . . *Xylaria (122)*

57(55') Interior of stroma zonate . . . . *Daldinia (120)*
57'(55') Interior of stroma not
zonate . . . . *Hypoxylon (118)*

58(3') Ascospores 2-celled 59
58'(3') Ascospores with more than
two cells 91

59(58) Ascospores hyaline or greenish 60
59'(58) Ascospores light to dark brown 70

60(59) Ascospores apiculate or
deeply constricted at septum 61

60'(59) Ascospores neither apiculate nor deeply constricted at septum 62

61(60) Ascospores apiculate at both ends . . . . *Hypomyces (144)*
61'(60) Ascospores deeply constricted at septum, cells subglobose, often separating . . . . *Hypocrea (142)*

62(60') Ascus unitunicate 66
62'(60') Ascus bitunicate 63

63(62') Ascoma perithecioid 64
63'(62') Ascoma elongate-stellate or dimidiate 65

64(63) Centrum containing pseudoparaphyses . . . . . *Didymella (146)*
64'(63) Centrum lacking pseudoparaphyses . . . . . . *Mycosphaerella (148)*

65(63') Ascoma dimidiate . . . . *Schizothyrium (154)*
65'(63') Ascoma elongate or stellate . . . . . *Glonium (152)*

66(62) Ascomata red or orange 67
66'(62) Ascomata brown to black 68

67(66) Perithecia immersed in a valsoid stroma . . . . *Cryphonectria (140)*
67'(66) Perithecia nonstromatic, or if stromatic, superficial on stroma or with only base immersed . . . . *Nectria (132)*

68(66') Perithecia immersed in a stroma outlined by a blackened zone . . . . *Diaporthe (136)*

68'(66') Perithecia immersed in host, but nonstromatic 69

69(68') Perithecia formed beneath a clypeus . . . . *Stegophora (130)*
69'(68') Perithecia lacking a clypeus . . . . *Gnomonia (128)*

70(59') Cells of ascospore different at maturity, one pigmented and one hyaline 71
70'(59') Cells of ascospore alike at maturity, both pigmented 73

71(70) Ascospore clavate when delimited . . . . *Podospora (160)*
71'(70) Ascospore cylindrical when delimited 72

72(71') Ascoma wall 5-layered . . . . . *Bombardia (156)*
72'(71') Ascoma wall 3-layered . . . . *Cercophora (158)*

73(70') Ascospores with germ slits . . . . *Delitschia (166)*
73'(70') Ascospores lacking germ slits 74

74(73') Ascomata immersed in host 75
74'(73') Ascomata superficial on host 76

75(74) Occurring in leaves . . . . *Venturia (164)*
75'(74) Occurring in herbaceous stems or twigs . . . . . . *Didymosphaeria (168)*

76(74') Ascostroma multi-loculate . . . . *Apiosporina (170)*

76'(74') Ascostroma uniloculate, dimidiate . . . . *Lembosia (172)*

77(1) Ascoma apothecioid 78
77'(1) Ascoma perithecioid or mussel-shaped 80

78(77) Ascoma erect, stipitate-spathulate . . . . *Spathularia (198)*
78'(77) Ascoma flat, immersed in host 79

79(78') Ascoma elliptical, opening by a single slit . . . . *Lophodermium (192)*
79'(78') Ascoma circular to oval, opening by several slits . . . . *Rhytisma (194)*

80(77') Ascoma erect, mussel-shaped . . . . *Lophium (196)*
80'(77') Ascoma perithecioid 81

81(80') Ascus unitunicate 82
81'(80') Ascus bitunicate . . . . *Cochliobolus (180)*

82(81) Perithecia immersed in a stroma 83
82'(81) Perithecia lacking a stroma 89

83(82) Perithecial stroma flat 84
83'(82) Perithecial stroma stipitate-capitate 87

84(83) Stroma encircling host stem 85
84'(83) Stroma flat on host leaves 86

85(84) Stroma yellow-orange . . . . *Epichloë (184)*
85'(84) Stroma brown to black . . . . *Balansia (182)*

86(84') Stroma broad and flat; ascus cylindrical with

a thick cap . . . . *Balansia (182)*

86'(84') Stroma narrow and linear; ascus widest in middle, with thin, dome-shaped cap . . . . *Myriogenospora (186)*

87(83') Perithecial stromata formed from sclerotia . . . . . . *Claviceps (188)*

87'(83') Perithecial stromata formed on host tissues 87

88(87') Stromata formed on grasses . . . . *Balansia (182)*

88'(87') Stromata formed on insects or other fungi . . . . *Cordyceps (190)*

89(82') Perithecia lying horizontally in host tissues . . . . *Linospora (176)*

89'(82') Perithecia arranged vertically in host 90

90(89') Occurring on grasses and sedges . . . *Gaeumannomyces (178)*

90'(89') Occurring on palms and dicotyledons . . . . *Ophiodothella (174)*

91(58') Ascospores phragmosporous 92

91'(58') Ascospores dictyosporous 105

92(91) Ascospores hyaline 93

92'(91) Ascospores yellowish to brown 101

93(92) Ascoma stipitate-capitate . . . . *Leotia (68)*

93'(92) Ascoma a perithecium or ascostroma 94

94(93') Asci unitunicate 95

94'(93') Asci bitunicate 98

95(94) Ascoma brown . . . . *Gaeumannomyces (178)*
95'(94) Ascoma red, orange, blue
or violet 96

96(95') Ascoma blue to violet . . . . . *Gibberella (202)*
96'(95') Ascoma red to orange 97

97(96') Outer wall cells thick-walled and
pseudoparenchymatous; anamorph
Cylindrocladium . . . . *Calonectria (200)*
97'(96') Outer wall cells variable;
anamorph variable, but not
Cylindrocladium, or
lacking . . . . *Nectria (132)*

98(94') Ascomata formed on a black
subiculum on
host . . . . *Scorias (208)*
98'(94') Ascomata not on a subiculum 99

99(98') Ascoma a multiloculate ascostroma
with asci irregularly
scattered . . . . *Elsinoë (210)*
99'(98') Ascoma a perithecioid
pseudothecium 100

100(99') Centrum containing pseudoparaphyses . . . . . . *Setosphaeria (204)*
100'(99') Centrum lacking
pseudoparaphyses . . *Sphaerulina (206)*

101(92') Ascoma a stipitate-clavate
apothecium . . . . *Geoglossum (220)*
101'(92') Ascoma perithecioid or
hysterothecioid 102

102(101') Ascoma an elongate hysterothecium opening
by a slit . . . . *Hysterium (218)*

102'(101') Ascoma perithecioid 103

103(102') Ascospores or part-spores with a germ slit . . . . *Sporormiella (216)*

103'(102') Ascospores lacking germ slits 104

104(103') Ascoma superficial on host, ascospores dark brown . . . . *Meliola (212)*

104'(103') Ascoma immersed in host, ascospores yellowish to light brown . . . *Leptosphaeria (214)*

105(91') Ascospores hyaline 106

105'(91') Ascospores brown 107

106(105) Ascoma multiloculate with asci scattered throughout . . . . *Myriangium (222)*

106'(105) Ascoma uniloculate, with asci in a fascicle . . . . . *Leptosphaerulina (224)*

107(105') Ascoma an elongate hysterothecium opening by a slit . . . *Hysterographium (230)*

107'(105') Ascoma perithecioid 108

108(107') Ascospores usually longer than 50μm, with 7 or fewer septa . . . . *Pyrenophora (228)*

108'(107') Ascospores usually shorter than 50μm, but those longer with more than 7 septa . . . . *Pleospora (226)*

*Numbers in parentheses are page numbers of descriptions.

DESCRIPTIONS AND ILLUSTRATIONS OF GENERA

CERATOCYSTIS Ellis & Halst.

Ascomata single to aggregated, superficial or immersed in substrate, ostiolate or non-ostiolate, globose, subglobose or obpyriform, hyaline to dark brown or black; ostiolar necks when present usually very long and cylindrical, often with a ring of hyphae around ostiole. Centrum containing pseudoparenchyma cells. Asci globose to subglobose, produced at different levels in the centrum, 8-spored, evanescent at maturity. Ascospores hyaline, 1-celled, with or without a gelatinous sheath, variable in shape, curved, lunate, reniform, or hat-shaped. (Fig. 3A,D).

Anamorphs: *Chalara, Gabarnaudia, Graphilbum, Graphiocladiella, Graphium, Hyalodendron, Hyalopesotum, Hyalorhinocladiella, Leptographium, Pachnodium, Pesotum, Phialocephala, Phialographium, Sporothrix,* and *Verticicladiella.*

Habitat: On living or dead stems or roots of vascular plants; occasionally on other substrates. Several species are associated with bark beetles.

Representative species: *Ceratocystis fimbriata* Ellis & Halst. (Anam. *Chalara* sp.), cause of black rot of sweet potato and other crops; *C. pilifera* (Fr.:Fr.) C. Moreau [=*Ophiostoma piliferum* (Fr.:Fr.) Syd. & P. Syd.] (Anam. *Hyalodendron* sp.), cause of blue stain of lumber, and *C. ulmi* (Buisman) C. Moreau [=*O. ulmi* (Buisman) Nannf.][Anam. *Pesotum ulmi* (M. B. Schwartz) Crane & Schoknecht], cause of Dutch elm disease.

Comments: There continues to be much disagreement over the delimitation of genera in this complex of fungi. Non-ostiolate species have been placed in *Europhium* Parker. De Hoog and Scheffer separate *Ceratocystis* and *Ophiostoma* as follows: in *Ceratocystis* the anamorph is endoconidial (*Chalara*), the cell walls lack cellulose and rhamnose, and there is no growth on cyclohexamide; in

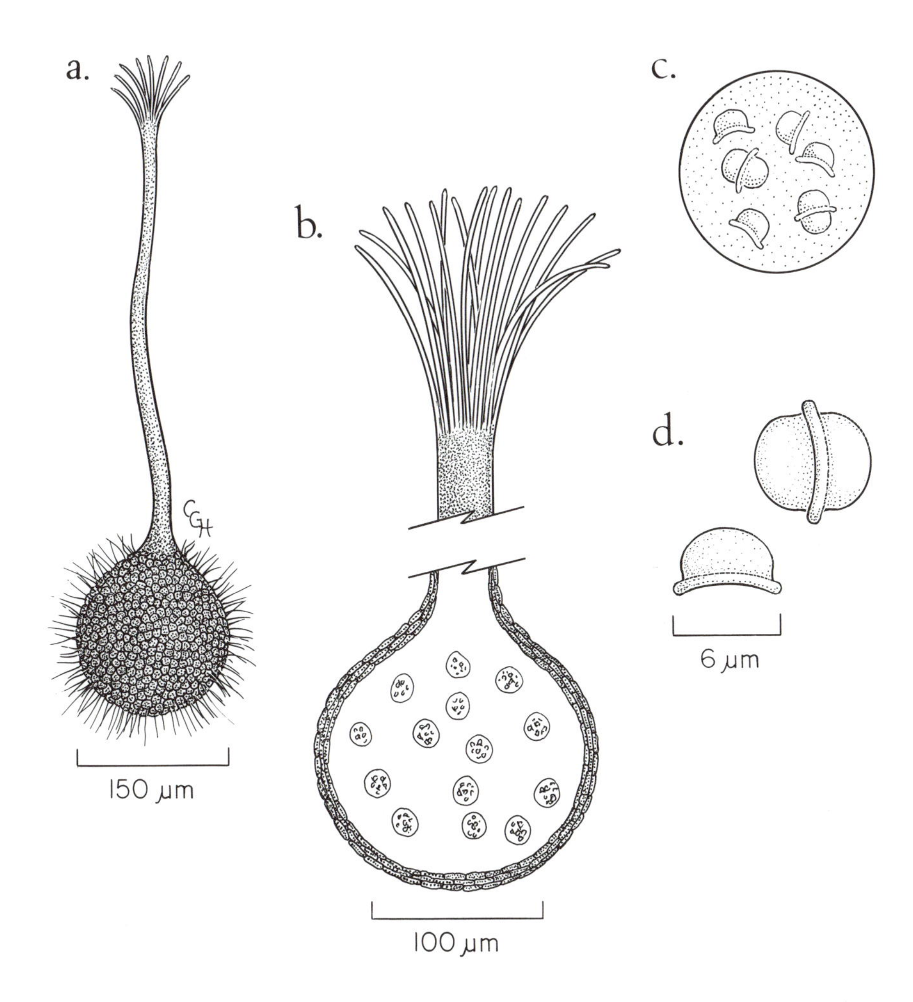

Ceratocystis fimbriata. **a.** Mature ascoma. **b**. Bottom, section through venter with scattered, globose asci; top, detail of apex of ostiolar neck. **c**. Ascus with ascospores. **d**. Ascospores.

Ophiostoma Syd. & P. Syd. the anamorphs are in genera other than *Chalara*, the cell walls contain cellulose and rhamnose, and there is growth on cyclohexamide. Species with elongate, falcate ascospores with attenuated ends formed in fusiform or clavate asci are placed in *Ceratocystiopsis* Upadhyay & Kendrick. In *Sphaeronaemella* Karsten, the light-colored ascomata contain brown ascospores with germ slits.

References: Domsch, et al., 1980; Ellis and Ellis, 1985; Griffin, 1966; Harrington, 1981; Hoog and Scheffer, 1984; Hunt, 1956; Matsushima, 1975; Olchowecki and Reid, 1974; Parker, 1957; Upadhyay, 1981; Weijman and de Hoog, 1975; CMI 141, 142, 143, 361.

→

Figure 1. **A-C.** *Brasiliomyces malachrae*. **A.** Optical section through ascoma with asci and ascospores. X661. **B.** Ascus with five ascospores. X682. **C.** Mature ascospore. X696. *Blumeria graminis*. **D.** Ascomata surrounded by hyphae of secondary mycelium. X41. **E.** Top view of ascoma with curved hyphae of secondary mycelium. X127. **F.** Immature ascus. X490. **G.** Side view of ascoma of *Podosphaera tridactyla*. X229. **H.** Ascomal appendage of *Uncinula parvula*. X1284. **I.** Mature ascospore of *Microsphaera ravenelii*. X1070.

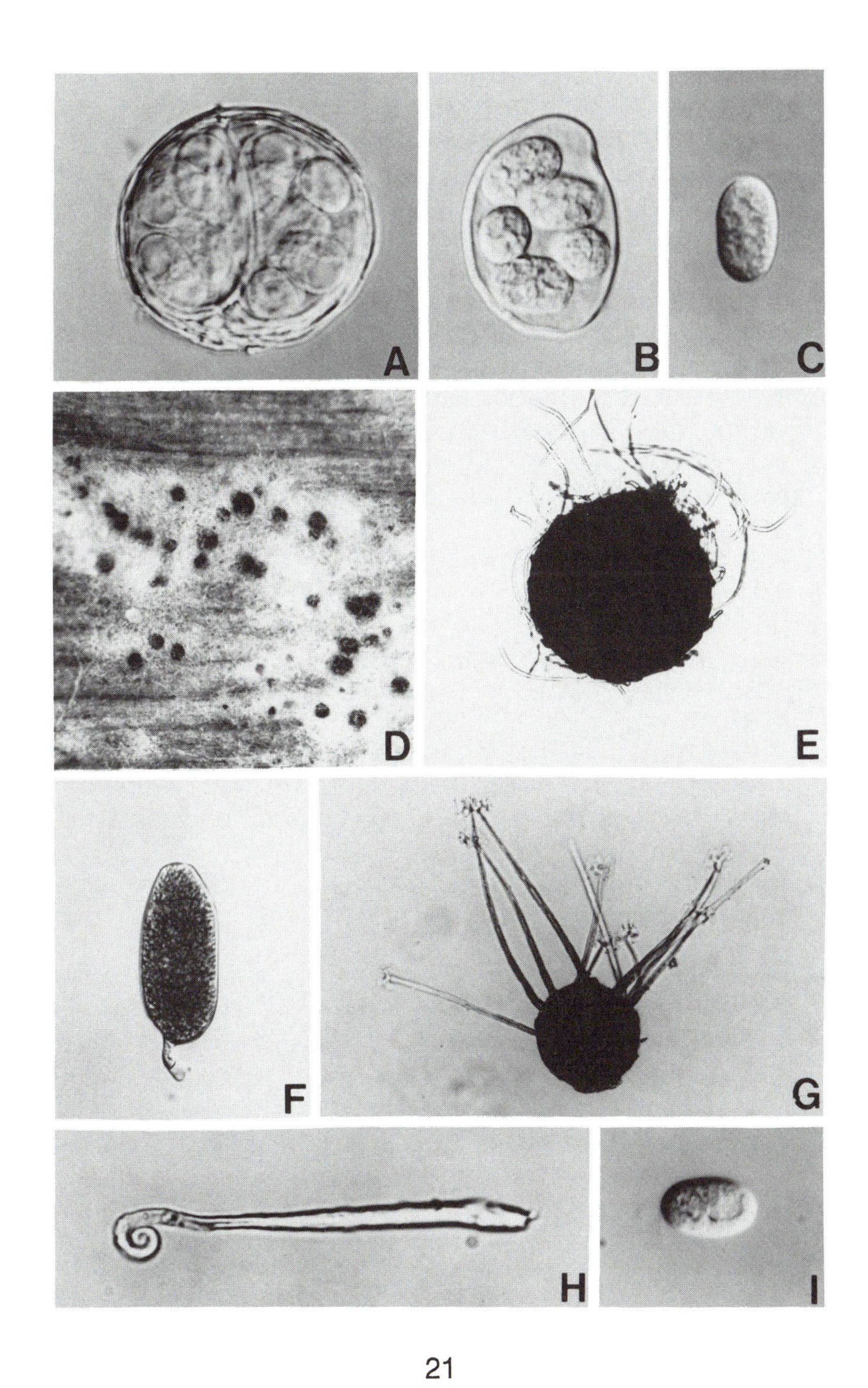
A
B
C
D
E
F
G
H
I

BRASILIOMYCES **Viégas**

Mycelium superficial, hyaline, forming white colonies on host tissues, bearing erect, tapered, hyaline setae in one species. Ascomata globose to subglobose, nonostiolate, seated on surface of mycelium; appendages lacking, but basal attachment hyphae often present. Ascomal wall thin, consisting of a single layer of undifferentiated cells, hyaline to sometimes yellowish or light brown with age. Asci unitunicate, saccate, occurring in fascicles of 2-5 per ascoma, each ascus containing 3-8 ascospores. Ascospores hyaline, one-celled, oblong to oval, with rounded ends. (Fig. 1A-C).

Anamorph: *Oidium*.

Habitat: On living leaves of angiosperms in the tropics and subtropics.

Representative species: *Brasiliomyces malachrae* (Seaver) Boesewinkel, cause of powdery mildew disease of *Gossypium* spp.

Comments: *Californiomyces* Braun and *Salmonia* Blumer & Müller are regarded as synonyms of *Brasiliomyces.*

References: Braun, 1987; Hanlin and Tortolero, 1984; Hodges, 1985; Viégas, 1944; Zheng, 1984.

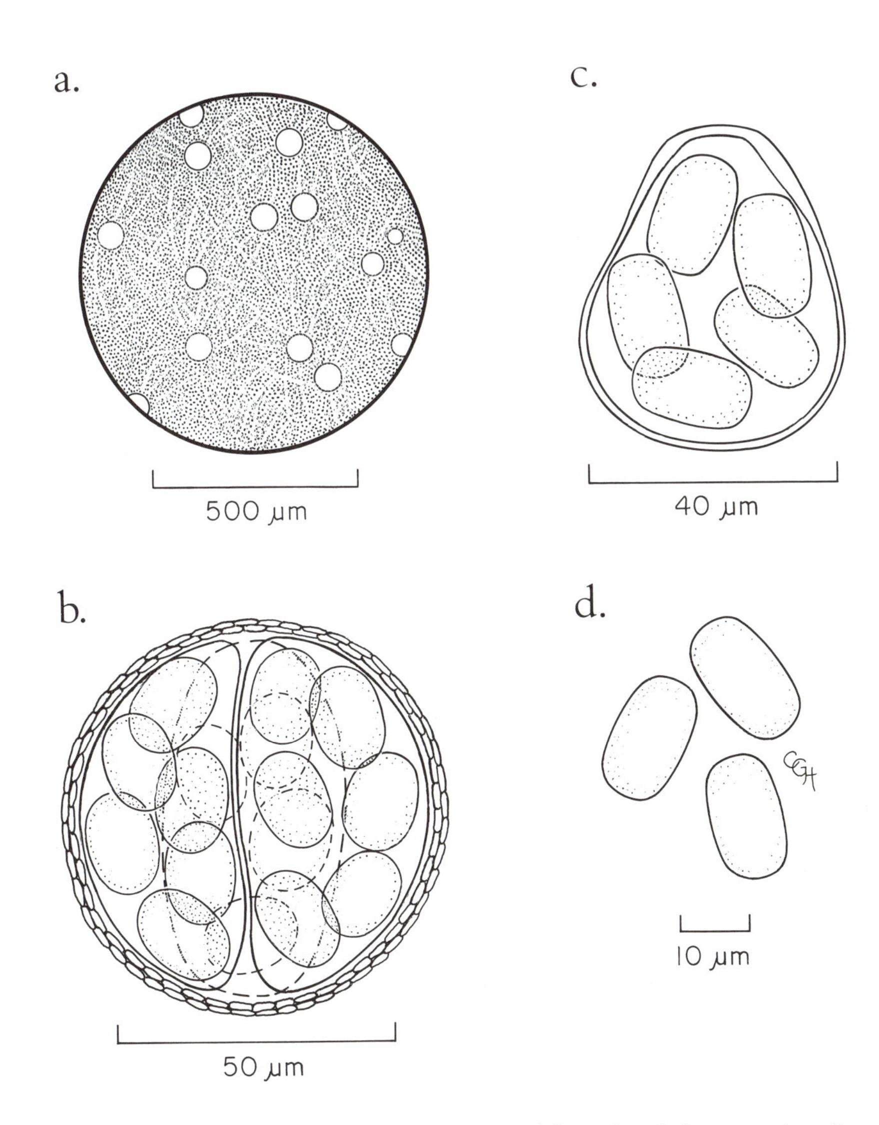

Brasiliomyces malachrae. **a.** Habit of globose, hyaline ascomata on cotton leaf. **b.** Optical section through ascoma with asci and ascospores. **c.** Ascus with ascospores. **d.** Mature ascospores.

BLUMERIA Golovin ex E. O. Speer

Mycelium hyaline, superficial, forming white colonies on host tissues, producing abundant conidia. Haustoria with finger-like lobes. Older mycelium becoming light brown and forming numerous erect, curved, hyaline, thick-walled setae (secondary mycelium). Ascomata nonostiolate, formed among cells of secondary mycelium, circular in face view, somewhat flattened in side view, with a basal ring of mycelioid appendages, appendages sometimes becoming brownish. Ascomal wall thick, composed of several layers of pseudoparenchyma cells; outermost cells thick-walled and dark brown, inner cells hyaline and thin-walled. Asci several, unitunicate, clavate to subcylindrical or ellipsoid, with a short stalk, usually 8- spored, rarely 4-spored. Ascospores ellipsoid, one-celled, hyaline. (Fig. 1D-F).

Anamorph: *Oidium monilioides (Nees)* Link.

Habitat: On living leaves and stems of grasses.

Representative species: *Blumeria graminis* (DC.) E. O. Speer, cause of powdery mildew disease of grasses.

Comments: This is the only powdery mildew that occurs on monocotyledons. It has long been known as *Erysiphe graminis* DC., but it differs from all other powdery mildews in the formation of a setose secondary mycelium, the digitate haustorium, and the restricted host range. The anamorph is also unique in being the only species of *Oidium* with a swollen basal cell.

References: Blumer, 1967; Braun, 1987; Ellis and Ellis, 1985; Homma, 1937; Junell, 1967; Parmalee, 1977; Sandu-Ville, 1967; Speer, 1973; CMI 153; FC 71; IM 41:C792.

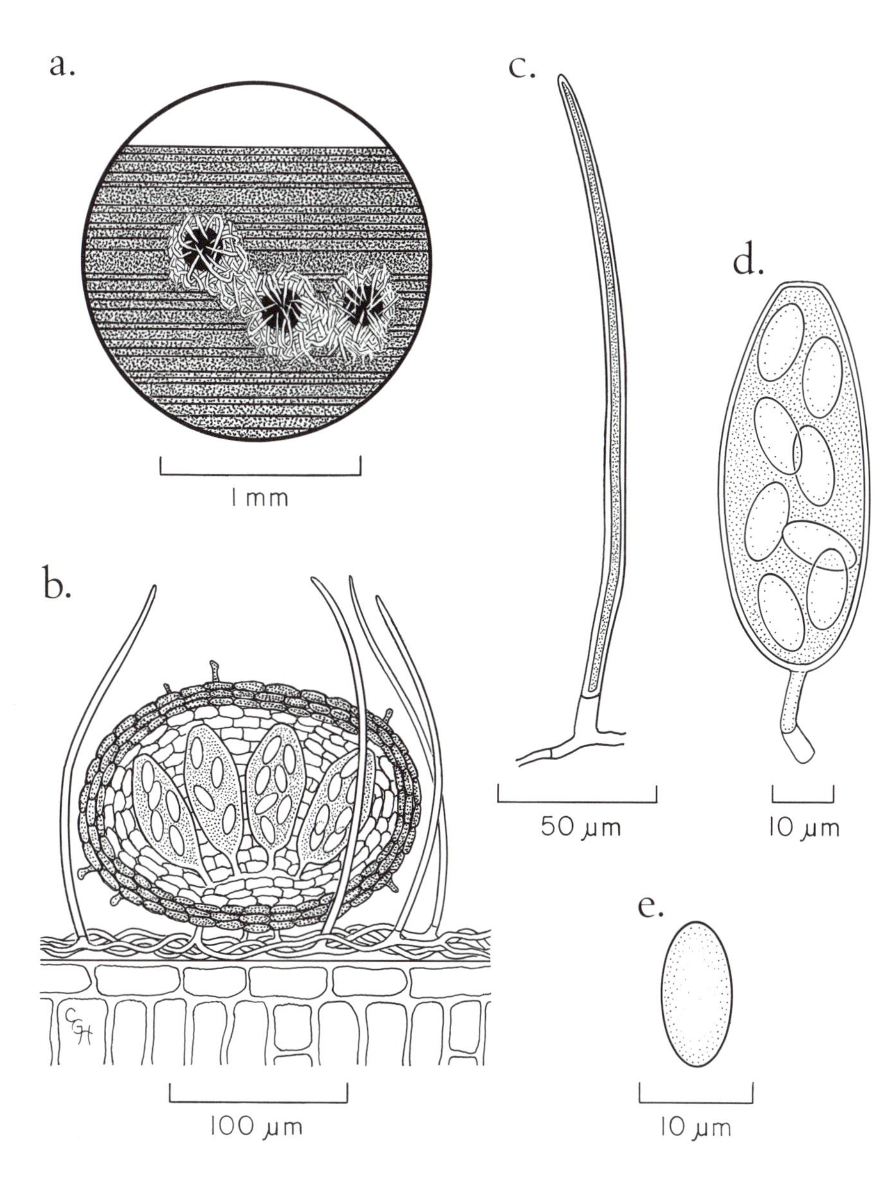

Blumeria graminis. **a.** Habit of perithecia on host leaf surrounded by "secondary mycelium". **b.** Section through perithecium on host leaf showing asci with ascospores and secondary hyphae. **c.** Setose hypha (secondary mycelium). **d.** Ascus with ascospores. **e.** Mature ascospore.

ERYSIPHE R. Hedw. ex DC.: Fr.

Mycelium hyaline, superficial, forming white colonies on host tissues, producing abundant conidia. Ascomata nonostiolate, dark brown, circular in face view, somewhat flattened in side view, with a basal ring of mycelioid appendages. Appendages hyaline, sometimes becoming brown near base. Ascomal wall thick, composed of pseudoparenchyma cells, outermost cells dark brown and thick-walled, inner cells hyaline and thin-walled. Ascoma containing many asci. Asci unitunicate, narrowly to broadly ellipsoid, to subclavate, short-stalked, 2-8-spored. Ascospores hyaline, 1-celled, ellipsoid to oblong.

Anamorph: *Oidium*.

Habitat: Parasitic on leaves, stems, flowers, and fruits of living herbaceous plants.

Representative species: *Erysiphe cichoracearum* DC. causing powdery mildew disease of a wide range of herbaceous plants, e.g., *Phlox* and *Zinnia*.

Comments: *Blumeria* Golovin ex E. O. Speer differs in the presence of setose secondary mycelium and in the gramineous host.

References: Blumer, 1967; Braun, 1987; Ellis and Ellis, 1985; Homma, 1937; Junell, 1967; Parmalee, 1977; Sandu-Ville, 1967; Zheng, 1981; CMI 151, 152, 154, 155, 156, 251, 509.

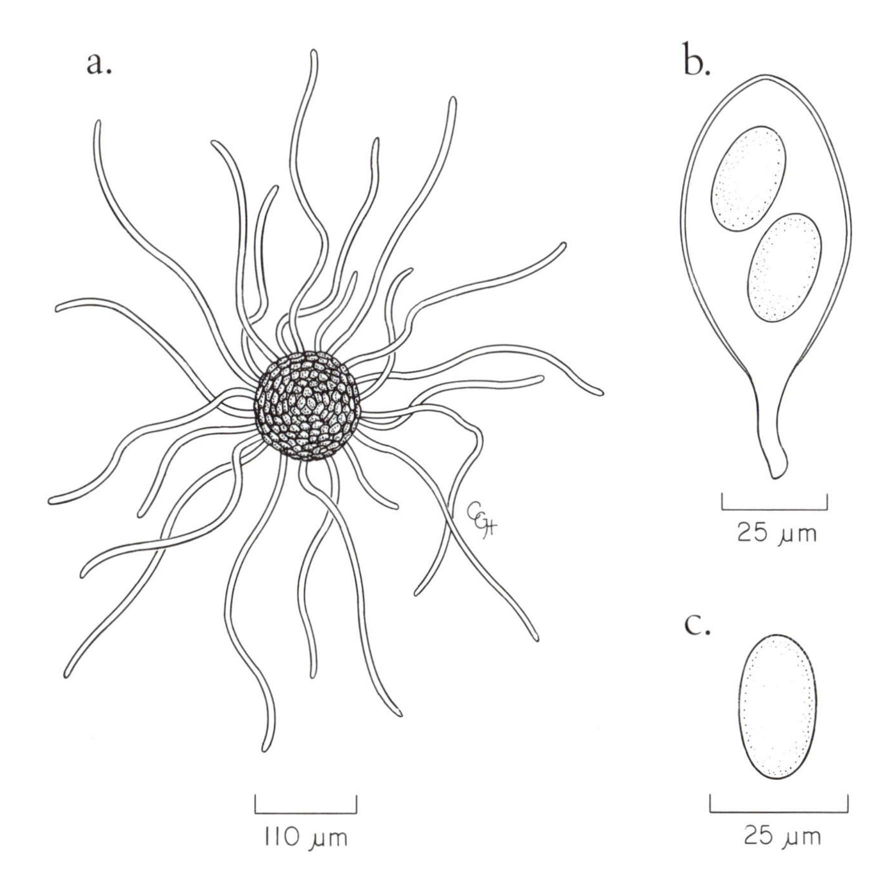

Erysiphe cichoracearum. **a.** Ascoma with appendages. **b.** Ascus and ascospores. **c.** Mature ascospore.

MICROSPHAERA Lév.

Mycelium hyaline, superficial, forming white colonies on host tissues, usually producing abundant conidia. Ascomata nonostiolate, circular in face view, somewhat flattened in side view, dark brown, with a basal ring of appendages that are dichotomously branched at the apex; appendages hyaline, but sometimes becoming light brown near base. Ascomal wall thick, composed of several layers of pseudoparenchyma cells, the outer cells thick-walled and dark brown, the inner cells hyaline and thin-walled. Asci several, unitunicate, subglobose to ovate or broadly ellipsoid, short-stalked, containing 3-8 ascospores. Ascospores hyaline, 1-celled, oblong-ellipsoid. (Fig. 1I, 2A-C).

Anamorph: *Oidium*.

Habitat: On living leaves, especially of trees and shrubs.

Representative species: *Microsphaera diffusa* Cooke & Peck on legumes and *M. penicillata* (Wallr.:Fr.) Lév. [=*M. alni* (Wallr.:Fr.) G. Wint.] on trees, shrubs, and herbaceous legumes.

Comments: *Podosphaera* Kunze differs in having a single ascus per ascoma.

References: Blumer, 1967; Braun, 1987; Ellis and Ellis, 1985; Homma, 1937; Junell, 1967; Parmalee, 1977; Sandu-Ville, 1967; CMI 183, 252.

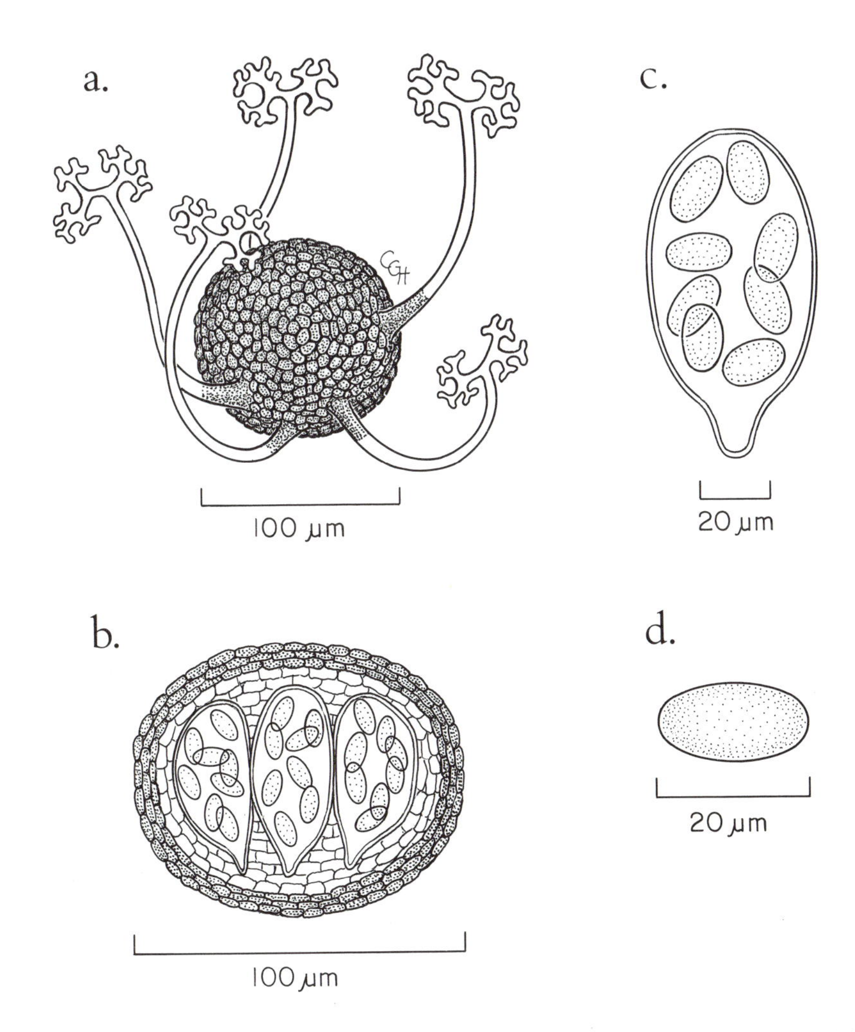

Microsphaera penicillata. **a.** Mature perithecium with appendages. **b.** Section through perithecium showing asci and ascospores. **c.** Ascus with ascospores. **d.** Mature ascospore.

PODOSPHAERA **Kunze**

Mycelium hyaline, superficial, forming white colonies on host tissues, producing abundant conidia. Ascomata nonostiolate, circular in face view, subglobose to somewhat flattened in side view, dark brown, with characteristic appendages. Appendages equatorially or apically inserted, dichotomously branched at the apex, hyaline, but becoming brownish near base. Ascomal wall thick, composed of pseudoparenchyma cells; outermost cells dark brown and thick-walled, inner cells hyaline and thin-walled. Ascoma containing a single ascus. Ascus unitunicate, broadly oval, broadly ellipsoid, to globose, containing 6-8 ascospores. Ascospores hyaline, 1-celled, ellipsoid to oblong. (Fig. 1G).

Anamorph: *Oidium.*

Habitat: Parasitic on leaves, flowers and stems of living woody shrubs and trees.

Representative species: *Podosphaera leucotricha* (Ellis & Everh.) E. S. Salmon (Anam. *Oidium farinosum* Cooke), cause of powdery mildew disease of apple.

Comments: *Microsphaera* Lév. differs in having several asci per ascoma.

References: Blumer, 1967; Braun, 1987; Ellis and Ellis, 1985; Ellis and Everhart, 1892; Homma, 1937; Junell, 1967; Parmalee, 1977; Sandu-Ville, 1967; CMI 158, 187; IM 1:C6.

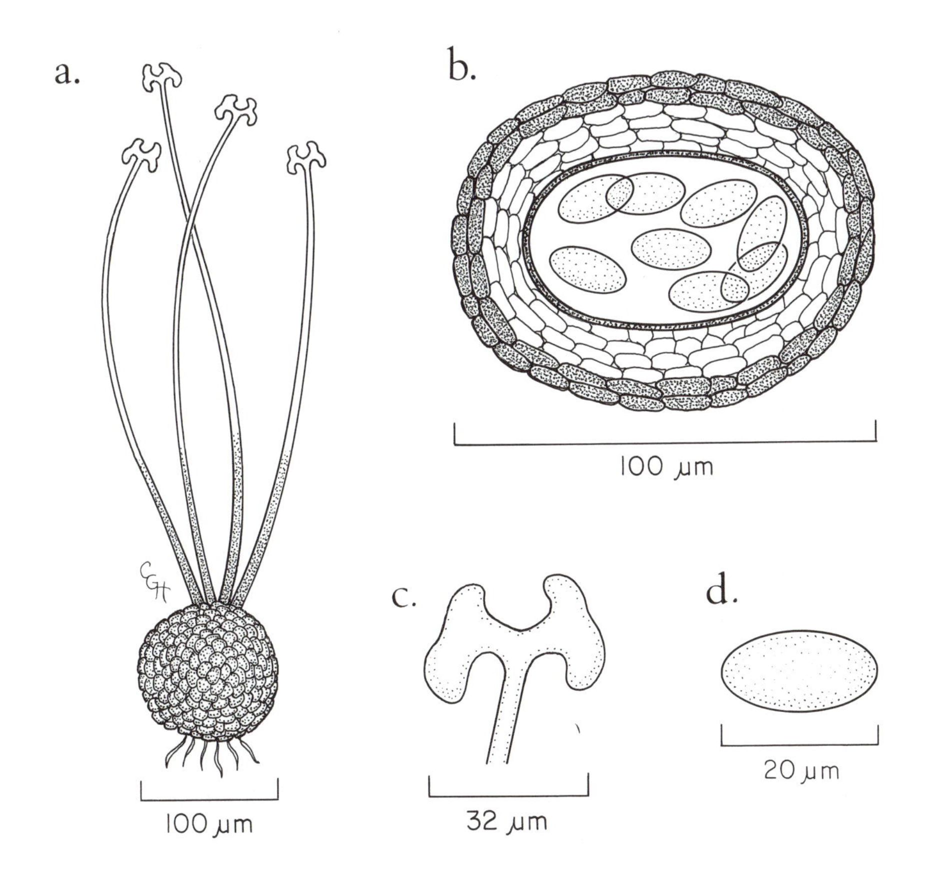

Podosphaera tridactyla. **a.** Perithecium with appendages. **b.** Section through perithecium showing single ascus and ascospores. **c.** Tip of appendage. **d.** Mature ascospore.

SPHAEROTHECA Lév.

Mycelium hyaline, superficial, forming white colonies on host, usually forming conidia. Ascomata nonostiolate, circular in face view, somewhat flattened in side view, dark brown, with a basal ring of hyaline, mycelioid appendages. Ascomal wall thick, composed of several layers of pseudoparenchyma cells, outermost cells thick-walled and dark brown, inner cells thin-walled and hyaline. Ascoma containing a single ascus; ascus unitunicate, broadly ellipsoidal to broadly oval or subglobose, 8-spored. Ascospores 1-celled, hyaline, ellipsoidal.

Anamorph: *Oidium.*

Habitat: On living leaves of herbaceous plants or shrubs.

Representative species: *Sphaerotheca pannosa* (Wallr. : Fr.) Lév. (Anam. *Oidium leucoconium* Desmaz.), cause of powdery mildew disease of rose.

Comments: *Erysiphe* R. Hedw. ex DC.:Fr. differs in having many asci.

References: Blumer, 1967; Homma, 1937; Junell, 1967; Parmalee, 1977; Sandu-Ville, 1967; CMI 159, 188; IM C6.

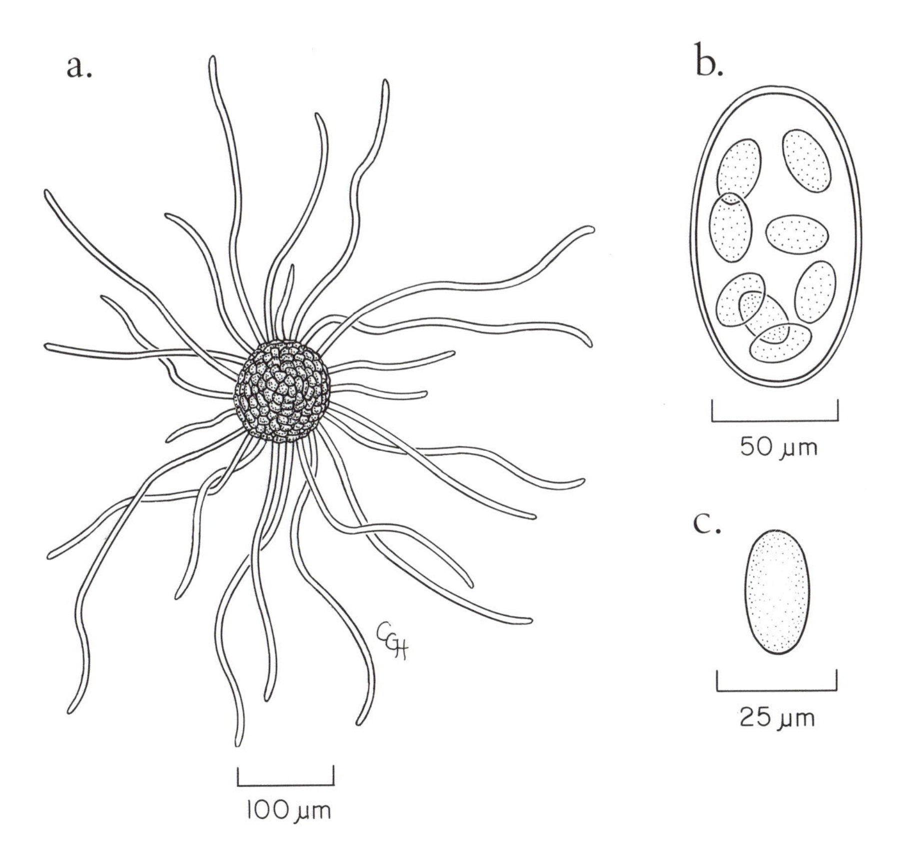

Sphaerotheca pannosa. **a.** Perithecium with appendages. **b.** Ascus with ascospores. **c.** Mature ascospore.

PHYLLACTINIA Lév.

Mycelium superficial and internal, hyaline, forming white colonies on host tissues, usually with conidium formation. Ascomata nonostiolate, circular in face view, somewhat flattened in side view, dark brown, upper surface covered with stiff, penicillate "brush cells", with a basal ring of stiff, tapering appendages with a bulbous base. Ascomal wall thick, composed of pseudoparenchyma cells, outermost cells thick-walled and dark brown, inner cells becoming thin-walled and hyaline toward the interior. Asci many, unitunicate, broadly oval to ellipsoid or broadly ellipsoid, short-stalked, containing 2-3 ascospores. Ascospores hyaline, 1-celled, oval. (Fig. 2D).

Anamorph: *Ovulariopsis*.

Habitat: Parasitic on living leaves of dicots, especially shrubs and woody plants.

Representative species: *Phyllactinia guttata* (Wallr.:Fr) Lév. on numerous trees and shrubs.

Comments: Both *Bulbouncinula* Zheng and *Queirozia* Viégas & Cardoso have appendages with a bulbous base, but the appendage apices are uncinate; *Bulbouncinula* also has short, capitate appendages and an *Oidium* anamorph, whereas *Queirozia* has a *Streptopodium (Ovulariopsis)* anamorph.

References: Blumer, 1967; Homma, 1937; Junell, 1967; Parmalee, 1977; Sandu-Ville, 1967; CMI 157, 186; IM C7.

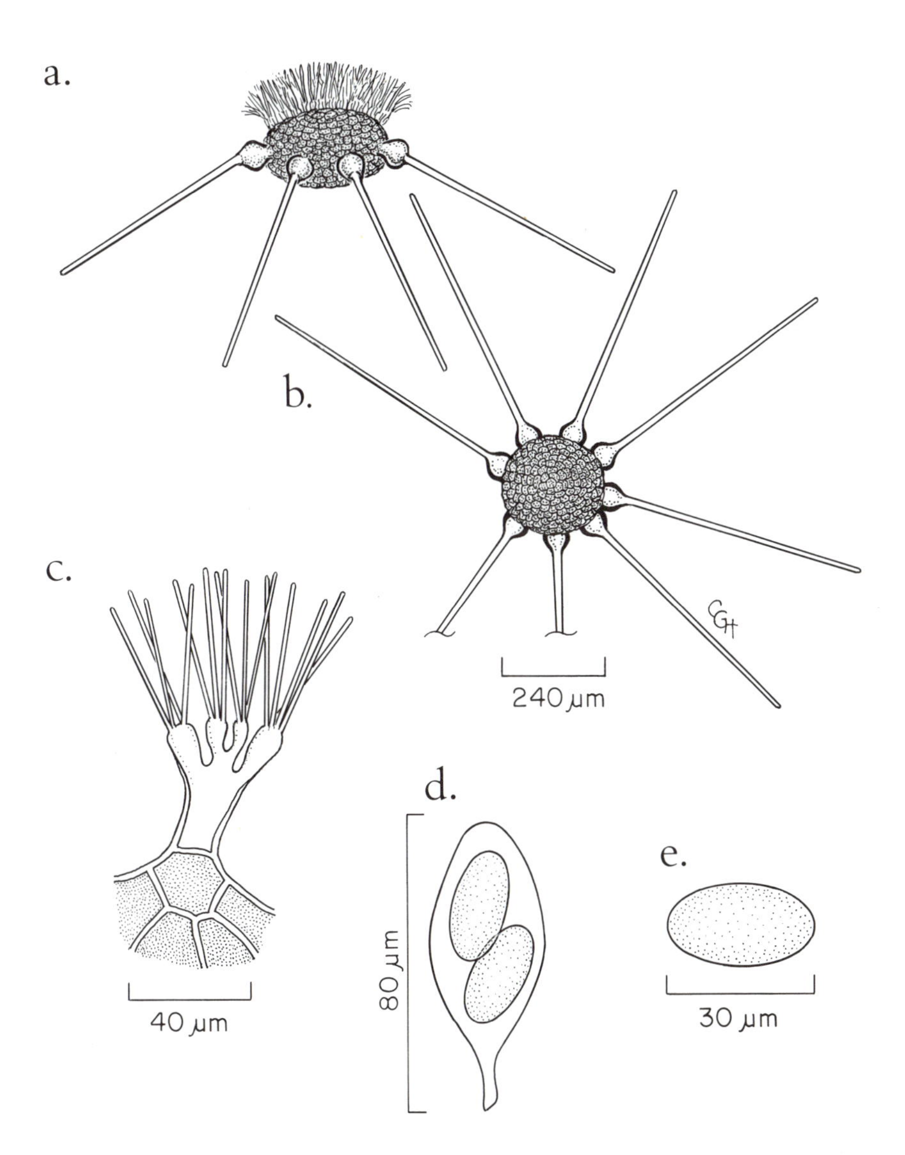

Phyllactinia guttata. **a.** Mature perithecium with penicillate "brush cells" on upper surface. Appendages have bent downward, lifting ascoma from surface of host. **b.** Top view of perithecium with appendages. **c.** Close-up of brush cell. **d.** Ascus with ascospores. **e.** Mature ascospore.

UNCINULA Lév.

Mycelium hyaline, superficial, forming white colonies on the host and producing conidia. Ascomata nonostiolate, circular in face view, somewhat flattened in side view, dark brown, with stiff, hyaline appendages that are recurved (uncinate) at the apex. Appendages inserted equatorially or apically. Ascomal wall thick, composed of several layers of pseudoparenchyma cells, the outermost cells thick-walled and dark brown, the inner cells hyaline and thin-walled. Asci several, unitunicate, narrowly to broadly ellipsoid, to narrowly oblong to obovate, short-stalked. Ascospores hyaline, 1-celled, oblong to ellipsoid. (Fig. 1H).

Anamorph: *Oidium.*

Habitat: Parasitic on leaves and stems of woody plants.

Representative species: *Uncinula necator* (Schwein.) Burrill (Anam. *Oidium tuckeri* Berk.), cause of powdery mildew disease of grape (*Vitis* spp.).

Comments: The genera *Bulbouncinula* Zheng and *Uncinuliella* Zheng & Chen differ from *Uncinula* in having two types of appendages. *Bulbouncinula* has long, uncinate appendages with a bulbous basal cell, and short, capitate appendages; *Uncinuliella* has long, uncinate appendages and short, sickle-shaped appendages. In *Sawadaia* Miyabe the appendages are often branched, with uncinate tips, whereas in *Furcouncinula* Z. X. Chen the appendage apices are uncinate, with forked tips. In *Pleochaeta* Sacc. & Speg. and *Queirozia* Viégas & Cardoso the ascomata are turbinate in side view, with a broad layer of asci and apically attached appendages. In *Queirozia* the appendages have a bulbous base, which is lacking in *Pleochaeta. Pleochaeta* and *Queirozia* also have *Streptopodium* anamorphs.

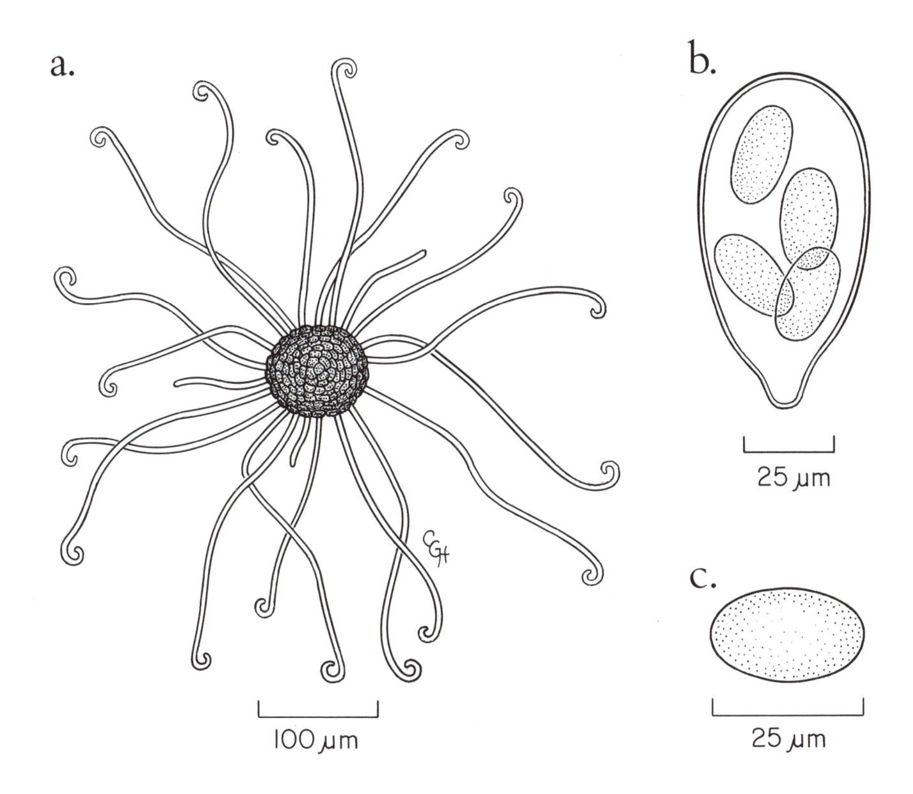

Uncinula necator. **a.** Perithecium with appendages. **b.** Ascus with ascospores. **c.** Mature ascospore.

References: Blumer, 1967; Braun, 1987; Ellis and Ellis, 1985; Ellis and Everhart, 1892; Homma, 1937; Junell, 1967; Parmalee, 1977; Sandu-Ville, 1967; Viégas, 1944; Zheng & Chen, 1977a,b, 1978, 1981; CMI 160, 190; IM 1:C8.

→

Fig. 2. **A.** *Microsphaera penicillata*. **A**. Section through ascoma with immature ascospores (left) and young ascus. X677. **B-C.** *Microsphaera ravenelii*. **B**. Asci with ascospores. X1531. **C**. Tip of ascomal appendage. X750. **D**. Top view of ascoma of *Phyllactinia angulata*. X97. **E-G**. *Ciboria carunculoides*. **E**. Apothecia arising from stromatized ovaries. X1.3. **F**. Ascus with ascospores. X650. **G**. Mature ascospores. X1312.

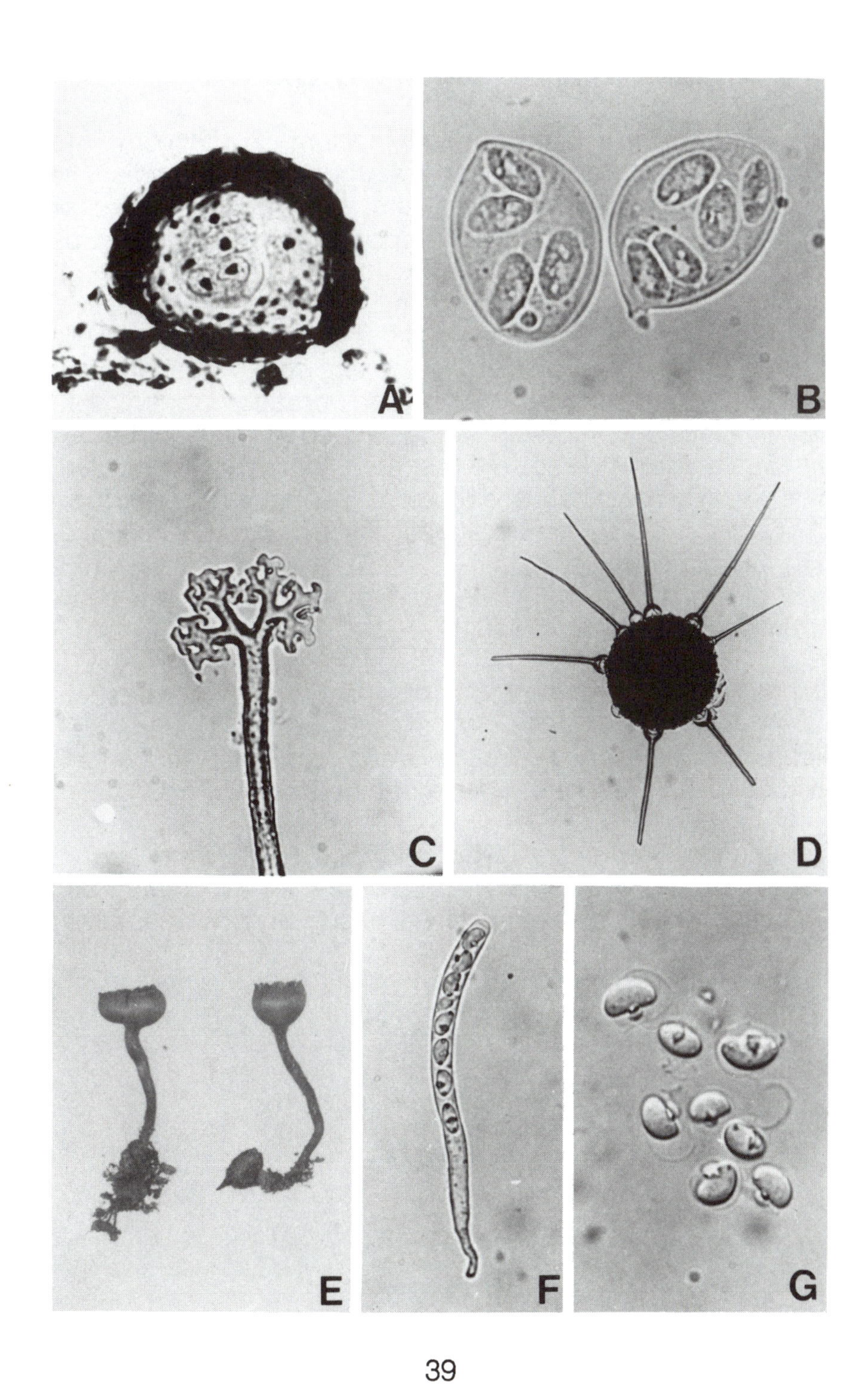
A
B
C
D
E
F
G

GLOMERELLA Spald. & H. Schrenk

Ascoma an ostiolate perithecium, obpyriform to subglobose, glabrous or with hairs around ostiole, partially or completely immersed in host tissue, single and scattered or aggregated, sometimes with a poorly developed clypeus around ostiolar necks; ostiolar neck inconspicuous or short, often paler than rest of ascoma, lined with periphyses. Ascomal wall pseudoparenchymatous, outer cells thicker-walled and pigmented, inner cells flattened and hyaline. Asci unitunicate, thin-walled, broadly cylindrical to slightly clavate or ellipsoid, with rounded, nonamyloid apex, sessile or short-stalked, 4-6-8-spored. Ascospores 1-celled, hyaline, biseriate in ascus, ellipsoidal or subcylindrical, straight or curved, often inaequilateral, less than 20 µm long. (Fig. 3B-C,E).

Anamorph: *Colletotrichum.*

Habitat: Parasites or saprobes on vascular plant tissues.

Representative species: *Glomerella cingulata* (Stoneman) Spald. & H. Schrenk, cause of bitter rot of apple.

Comments: *Glomerella* is distinguished from *Physalospora* Niessl by the smaller size of the ascomata and ascospores, the usually clavate ascus, and the anamorph.

References: Arx and Müller, 1954; Dennis, 1981; Munk, 1957; Viégas, 1944; CMI 133, 315.

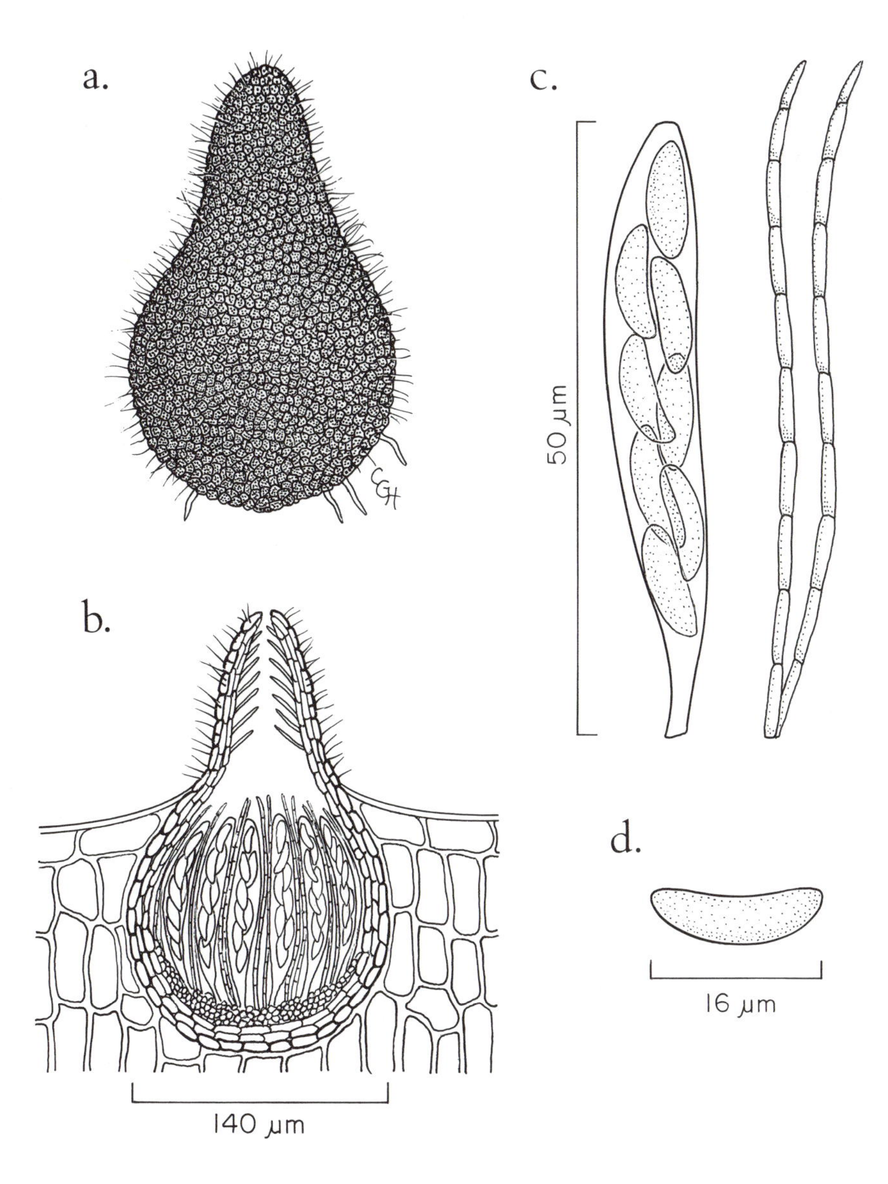

Glomerella cingulata. **a.** Mature perithecium. **b.** Section through perithecium in leaf, with paraphyses, asci and ascospores, and ostiolar neck lined with periphyses. **c.** Paraphyses and asci with ascospores. **d.** Mature ascospore.

PHYLLACHORA Nitschke ex Fuckel

Ascoma an ostiolate, obpyriform to subglobose perithecium, immersed in host tissue, with ostiole becoming erumpent through host epidermis, occurring singly but frequently crowded together, surrounded at least around neck by a dark clypeus, clypeus often amphigenous. Ascomal wall composed of flattened cells, outer cells dark brown, inner cells hyaline. Asci unitunicate, cylindrical to ellipsoid or occasionally saccate, apex undifferentiated, nonamyloid, containing 8 ascospores, variously arranged. Ascospores hyaline, 1-celled, oval to ovoid or globose.

Anamorph: *Leptostromella* known in one species, but many species form spermatia.

Habitat: Parasitic on leaves of vascular plants, especially common in tropical regions.

Representative species: *Phyllachora graminis* (Pers.: Fr.) Nitschke, cause of tar spot disease of grasses in temperate regions, and *P. maydis* Maubl., cause of tar spot of maize in the tropics.

Comments: *Glomerella* Spald. & H. Schrenk and *Physalospora* Niessl differ in lacking a distinct clypeus around the ostiolar necks. *Catacauma* Theiss. & Syd. and *Trabutiella* Theiss. & Syd. have been segregated from *Phyllachora* on the basis of the position of the ascoma in relation to the host tissues. In *Catacauma* the ascoma is subepidermal and in *Trabutiella* it is subcuticular, whereas in *Phyllachora* the ascoma forms in the palisade and mesophyll tissues. These genera are regarded as synonyms of *Phyllachora*.

References: Arx and Müller, 1954; Chardon and Toro, 1930; Dennis, 1981; Ellis and Ellis, 1985; Ellis and Everhart, 1892; Munk, 1957; Orton, 1944; Parbery, 1967; Seaver and Chardon, 1926; Stevens, 1924, 1927; Viégas, 1944.

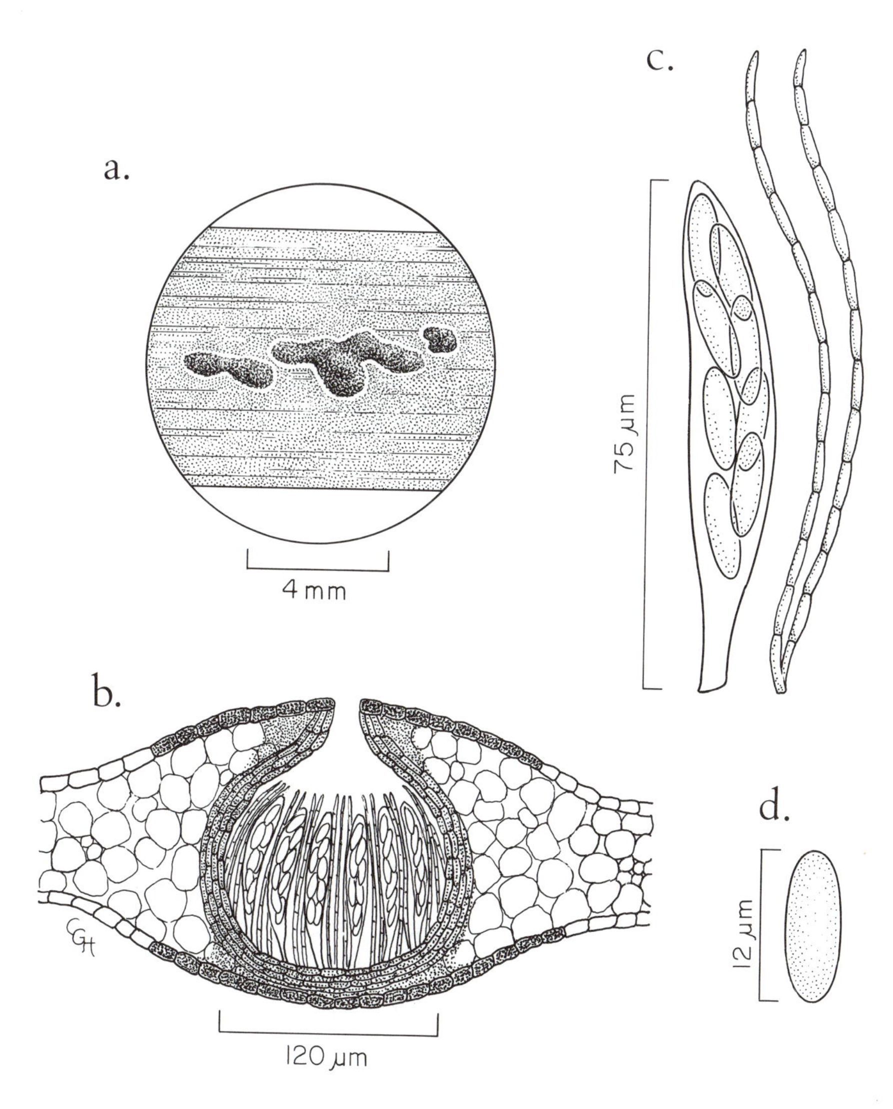

Phyllachora graminis. **a.** Erumpent black clypeus on grass leaf. **b.** Section through perithecium in host leaf, with paraphyses, and asci and ascospores. Note black clypeus in upper and lower epidermis. **c.** Paraphyses and ascus with ascospores. **d.** Mature ascospore.

PHYSALOSPORA Niessl

Ascoma an ostiolate perithecium, immersed in host tissues, with erumpent ostiolar neck; perithecia single, globose, obpyriform or vertically ellipsoidal, brown, glabrous or with hairs, often with short setae around ostiole. Ostiolar neck papillate or conical, often darker than rest of ascoma, lined with periphyses. Perithecial wall of flattened cells, outer cells subhyaline or olive-brown, inner cells hyaline. Paraphyses filiform, often with slightly clavate tips, sometimes with mucilaginous, coalesced walls. Asci unitunicate, clavate-cylindrical, truncate at apex, broadest in middle, tapering to a short stipe, some species with a simple, amyloid or nonamyloid apical ring, 8-spored. Ascospores 1-celled, hyaline or lightly colored, obovoid to elliptic, smooth or finely roughened, with granular contents, sometimes with a thin, gelatinous sheath.

Anamorph: None reported.

Habitat: On living or dead leaves of flowering plants, occasionally on twigs.

Representative species: *Physalospora alpestris* Niessl, on *Carex* spp.

Comments: *Physalospora* is distinguished from *Glomerella* Spald. & H. Schrenk by its larger ascomata and ascospores, the fusoid ascus, and the lack of an anamorph. The apple black rot pathogen, commonly referred to as *P. obtusa* (Schwein.) Cooke, belongs in *Botryosphaeria*.

References: Arx and Müller, 1954; Barr, 1970; Ellis and Everhart, 1892; Munk, 1957; IM 40:C778.

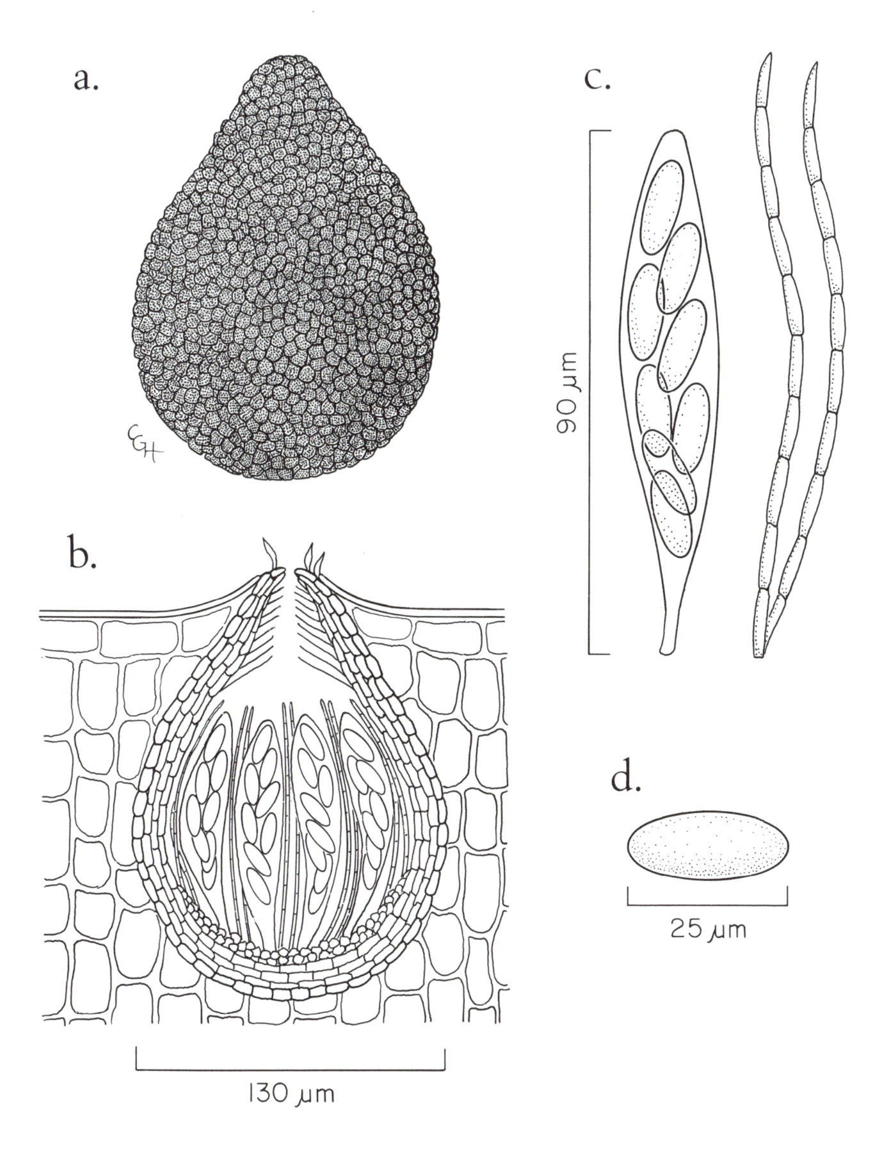

Physalospora alpestris. **a.** Exterior of perithecium. **b.** Section through perithecium in leaf, with paraphyses, asci and ascospores, and ostiolar neck lined with periphyses. **c.** Paraphyses, and ascus with ascospores. **d.** Mature ascospore.

GUIGNARDIA Viala & Ravaz

Ascoma a uniloculate, perithecioid pseudothecium, solitary to aggregated, immersed in host tissues or erumpent; pseudothecia dark brown, globose to subglobose, ostiolate, with ostiolar papilla or neck. Ascomal wall relatively thick, exterior composed of thick-walled pseudoparenchyma cells, interior cells thin-walled and hyaline. Centrum pseudoparenchymatous. Asci bitunicate, clavate to cylindrical, with short stalk, 8-spored. Ascospores hyaline, 1-celled, straight or curved, ovoid, ellipsoidal or rhomboidal, guttulate, usually widest in middle, with distinct mucilaginous appendages at one or both ends, less than 25 µm long.

Anamorph: *Phyllosticta (Phyllostictina)*; also a *Leptodothiorella* spermatial state.

Habitat: On gymnosperms, monocotyledons, and herbaceous dicotyledons, usually foliicolous.

Representative species: *Guignardia bidwellii* (Ellis) Viala & Ravaz [Anam. *Phyllosticta ampelicida* (Englem.) van der Aa], cause of black rot disease of grape (*Vitis* spp.) and related genera.

Comments: Barr considers *Guignardia* a synonym of *Botryosphaeria* Ces. & De Not., but most others separate the two genera. *Guignardia* differs from *Botryosphaeria* in having unilocular ascomata, smaller ascospores, and in the anamorph. Punithalingam restricts *Guignardia* to those species with *Phyllosticta* anamorphs; species with other anamorphs are placed in *Discosphaerina* Höhn. Bissett has proposed that *Guignardia* species with *Phyllosticta* anamorphs be placed in *Discochora* Höhn.

References: Arx and Müller, 1954; Barr, 1972; Bissett, 1986a,b; Ellis and Ellis, 1985; Punithalingam, 1974; Sivanesan, 1984; CMI 85, 467, 710; FC 21; IM 17:C248.

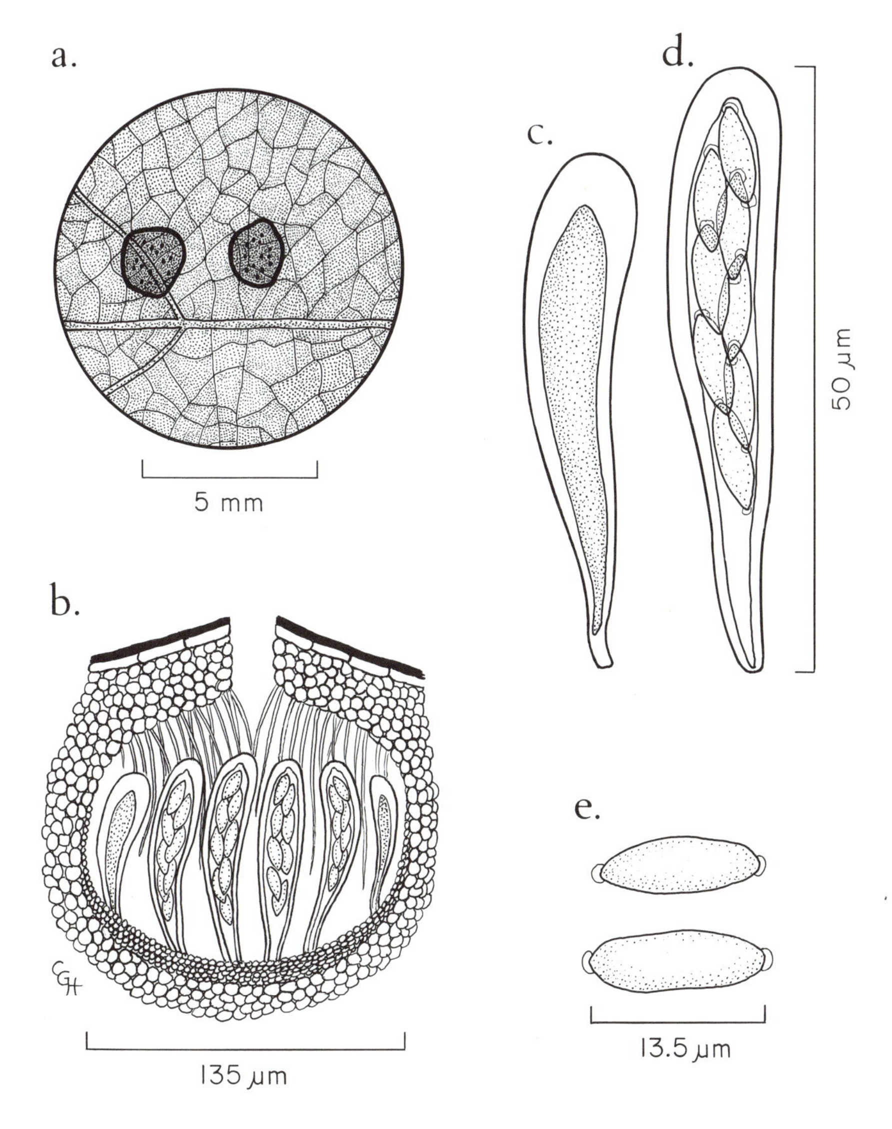

Guignardia bidwellii. **a.** Portion of leaf with lesions containing erumpent ostiolar necks (black dots) of ascomata. **b.** Section through mature pseudothecium, with asci and ascospores, and pseudoparaphyses. **c.** Young bitunicate ascus. **d.** Mature ascus with ascospores. **e.** Mature ascospores.

BOTRYOSPHAERIA Ces. & De Not.

Ascomata ascostromatic, black, usually multiloculate, sometimes uniloculate, ostiolate, with short neck, aggregated and erumpent through host tissues. Ascomal wall thick, pseudoparenchymatous, outer cells thick-walled and dark-brown, inner cells smaller and hyaline, with thin walls. Centrum containing numerous filamentous pseudoparaphyses. Asci bitunicate, clavate or oblong, stalked or sessile, 8-spored. Ascospores hyaline and 1-celled, but becoming brown and septate in some species, ovoid, fusoid to ellipsoid, often inaequilateral, usually widest in the middle, smooth, occasionally becoming verrucose, sometimes with a thin gelatinous coat. (Fig. 4C,E).

Anamorphs: *Botryodiplodia (Lasiodiplodia), Dothiorella, Diplodia, Macrophoma*, and *Sphaeropsis*.

Habitat: Usually occuring on woody or herbaceous stems.

Representative species: *Botryosphaeria rhodina* (Cooke) Arx (Anam. *Botryodiplodia theobromae* Pat.), associated with diseases of a wide variety of plants.

Comments: Species of *Botryosphaeria* are very common, usually occurring in the anamorphic state. There is much disagreement in the literature regarding both the delimitation of the genus and of species within the genus. Barr includes *Guignardia* Viala & Ravaz as a synonym of *Botryosphaeria*, but Sivanesan does not. *Botryosphaeria* has usually multiloculate ascostromata, larger ascospores, and different anamorphs than *Guignardia*. Also included as a synonym of *Botryosphaeria* is *Neodeightonia* Booth, which has brown ascospores and an unusual chambered ostiolar neck.

References: Arx and Müller, 1954; Barr, 1970, 1972; Dennis, 1981; Domsch et al., 1980; Ellis and Ellis, 1985; Ellis and Everhart, 1892; Sivanesan, 1984; Viégas, 1944; BK 377; CMI 394, 395, 774; FC 21; IM 12:C172.

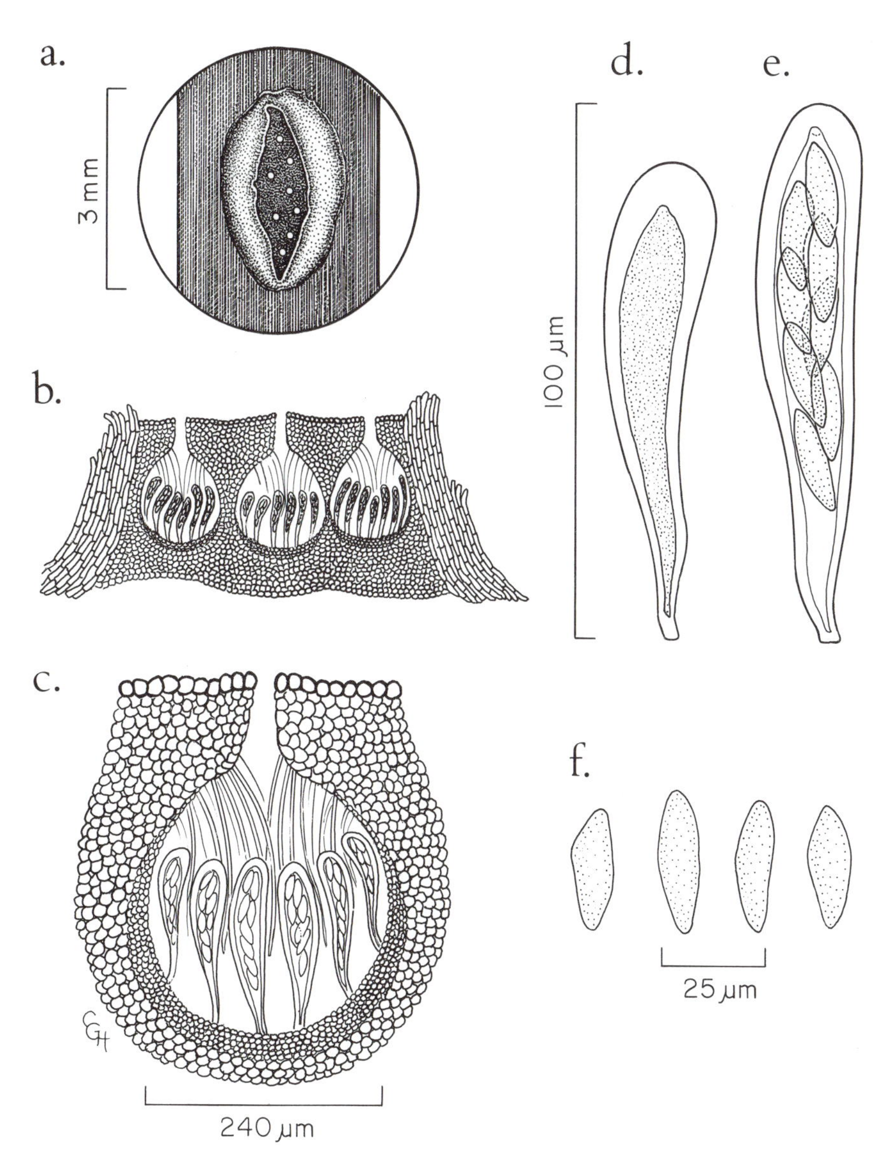

Botryosphaeria dothidea. **a.** Ascostroma erumpent through periderm of twig. **b.** Section through ascostroma with ascigerous locules containing asci; ascostroma is bordered by periderm of host. **c.** Close-up of ascigerous locule with asci and ascospores, and pseudoparaphyses. **d.** Young bitunicate ascus. **e.** Mature ascus with ascospores. **f.** Mature ascospores.

ALEURIA **Fuckel**

Ascoma a cupulate apothecium, small to large (up to 70 mm diameter), sessile to short-stipitate, scattered, gregarious or rarely caespitose. Receptacle deeply cupulate to saucer-shaped, outer surface paler than hymenium, smooth or mealy, in some species with hyaline, thin-walled, hair-like hyphae. Hymenium concave, smooth, orange-red, orange or yellow. Ectal excipulum pseudoparenchymatous, with cells mostly subglobose or polygonal-elongate. Medullary excipulum well differentiated, of textura intricata. Paraphyses simple, slender, enlarged at apex, straight, curved or hooked, with orange colored granular contents. Asci unitunicate, operculate, cylindrical, narrower towards base, not blueing in iodine, 8-spored. Ascospores hyaline, 1-celled, ellipsoidal or oblong ellipsoidal, with reticulate or verrucose wall ornamentation, containing one or two large oil droplets.

Anamorph: None reported.

Habitat: On damp soil and humus.

Representative species: *Aleuria aurantia* (Pers. ex Hook.) Fuckel, forming large orange apothecia on soil.

Comments: In *Leucoscypha* Boud. the hairs are well-developed and thick-walled and the hymenium is white to orange or red.

References: Dennis, 1981; Rifai, 1968; BK 98-99; IM 42:C804.

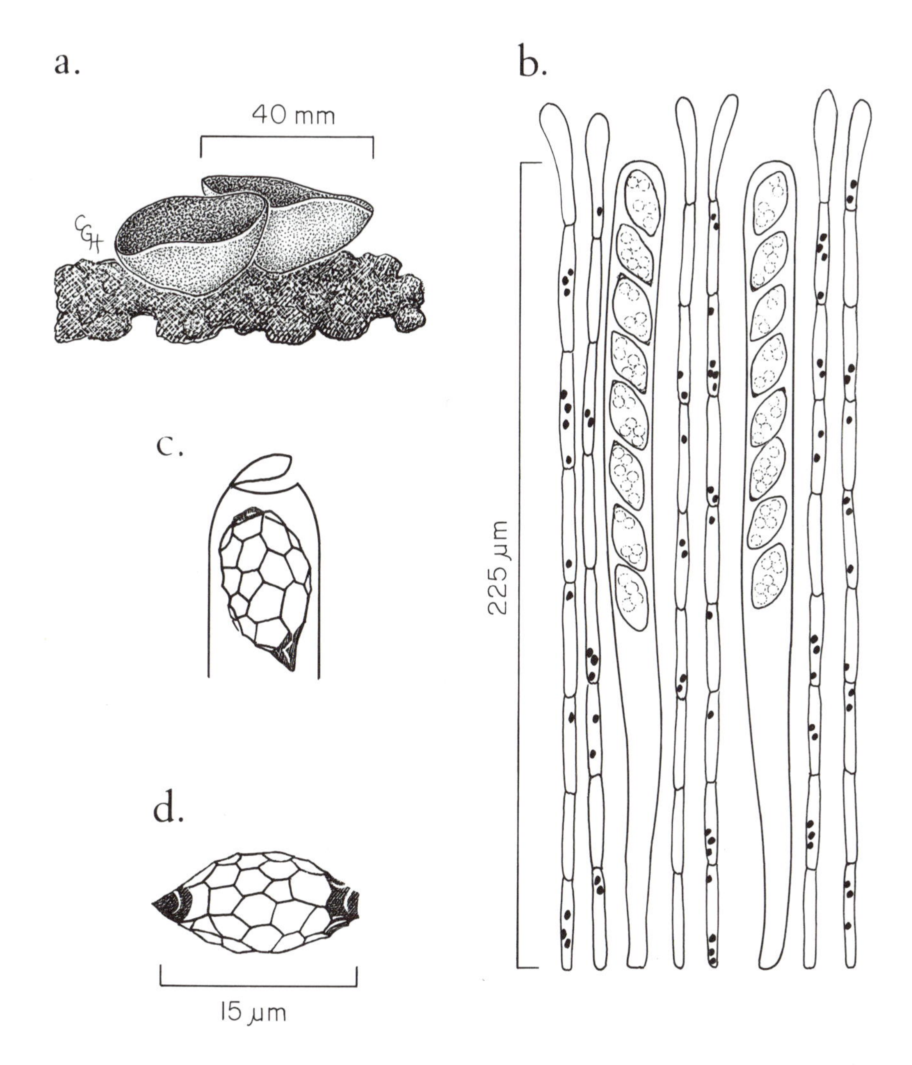

Aleuria aurantia. **a.** Mature apothecia on soil. **b.** Paraphyses with pigment granules and asci with ascospores. **c.** Ascus apex with operculum. **d.** Mature ascospore with reticulate wall ornamentation.

PEZIZA L.

Ascoma a cupulate apothecium, scattered to gregarious, sometimes caespitose, sessile or short-stalked, medium to large (up to 100 mm diam.), usually superficial, but some species immersed in soil when young. Disc smooth, deeply to shallowly concave, sometimes undulate or flattened, variously colored, white, brown, violet or other colors. Receptacle paler than disc, entire or crenate, smooth or downy, fleshy. Excipular tissues variable, prosenchymatous or pseudoparenchymatous, with large globose or polygonal cells. Paraphyses slender, septate, straight or curved, with apex slightly enlarged, often containing colored granules. Asci unitunicate, operculate, apex blueing in iodine, cylindrical, narrowing slightly towards base, 8-spored. Ascospores hyaline to subhyaline, 1-celled, ellipsoidal or rarely subglobose, uniseriate in ascus, smooth or ornamented.

Anamorphs: *Chromelosporium (Ostracoderma)* and *Oedocephalum*.

Habitat: On damp soil, rotting wood, burned areas, and some on dung.

Representative species: *Peziza ostracoderma* Korf [Anam. *Chromelosporium fulvum* (Link) McGinty, Henneb. & Korf], the peat mould, common in greenhouses.

Comments: Some authors place species with spherical spores in *Plicaria* Fuckel.

References: Dennis, 1981; Domsch et al., 1980; Ellis and Ellis, 1985; Rifai, 1968; BK 35-55; FC 168, 169.

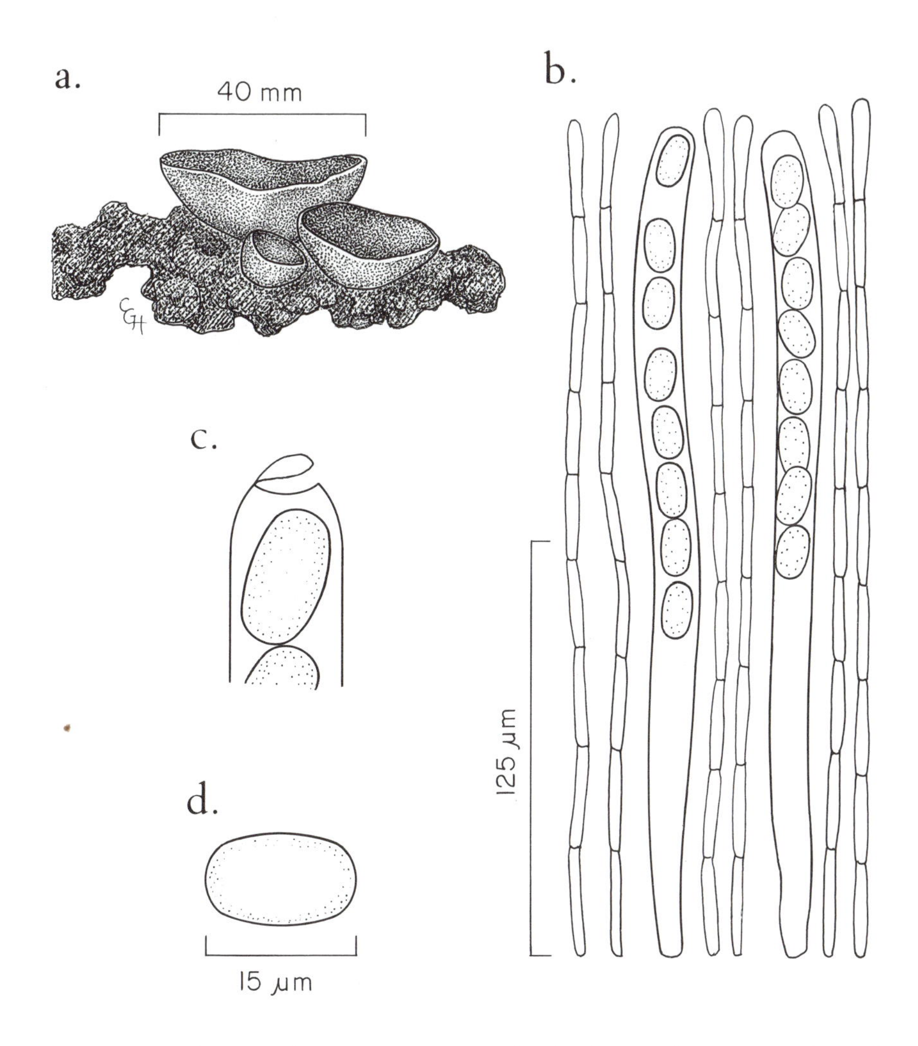

Peziza badia. **a.** Apothecia on soil. **b.** Paraphyses and asci with ascospores. **c.** Ascus apex with operculum. **d.** Mature ascospore.

SARCOSCYPHA (Fr.) Boud.

Ascoma an apothecium, saucer-shaped to cupulate, single to gregarious, sessile or stipitate, up to 5 cm in diameter. Hymenial surface concave, scarlet to orange or rarely whitish. Outer surface of receptacle even or crenulate, whitish, smooth, velvety or tomentose. Ectal excipulum thin, composed either of parallel prosenchymatous hyphal cells or of pseudoparenchyma tissue with angular or polygonal, slightly thick-walled cells (textura prismatica or textura angularis). Medullary excipulum of textura intricata, hyphae slender, septate, seldom branched. Paraphyses filiform, branched, septate, only slightly enlarged at apex. Asci unitunicate, suboperculate, long cylindrical, often tapering at base, thick-walled, not blueing in iodine, 8-spored. Ascospores 1-celled, hyaline, uniseriate in ascus, ellipsoidal to oblong ellipsoidal or subcylindrical, symmetrical, smooth, with oil globules.

Anamorph: None reported.

Habitat: On dead wood or soil.

Representative species: *Sarcoscypha coccinea* (Jacq. ex S. F. Gray) Lamb., with large scarlet apothecia.

Comments: Several authors have used *Plectania* Fuckel for this genus. *Microstoma* Bernstein differs in having a gelatinous layer in the apothecium. In *Phillipsia* Berk. the ascospores are longitudinally striate and flattened on one side.

References: Denison, 1972; Dennis, 1981; Ellis and Ellis, 1985; Kanouse, 1940; Rifai, 1968; BK 120; IM 28:C507.

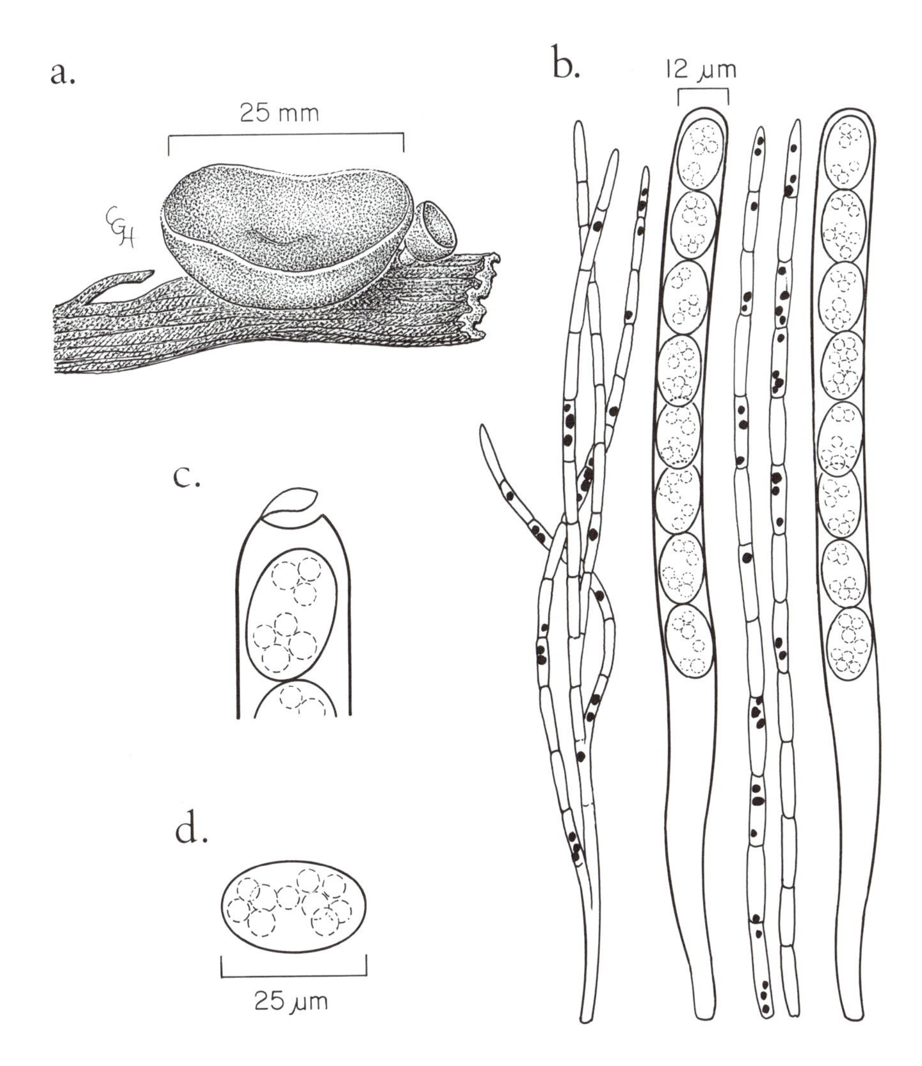

Sarcoscypha coccinea. **a.** Apothecia on wood. **b.** Paraphyses with pigment granules and asci with ascospores. **c.** Ascus apex with operculum. **d.** Mature ascospore with oil droplets.

URNULA Fr.

Ascoma an apothecium, solitary to clustered, up to 6 cm high; apothecium stipitate to occasionally sessile, dark-brown to black, deeply cupulate to urn-shaped, closed when young, opening at maturity, margin deeply stellate or notched, exterior tomentose. Hymenium brownish-black; paraphyses filiform, septate, often branched, slightly enlarged at tips; medullary excipulum long-celled, often gelatinous. Asci unitunicate, suboperculate, not blueing in iodine, cylindric to subcylindric, with a long, tapering stipe, 8-spored. Ascospores 1-celled, uniseriate in ascus, ovoid, ellipsoid to fusoid, smooth, hyaline.

Anamorph: *Conoplea* (*Strumella*).

Habitat: On wood in soil.

Representative species: *Urnula craterium* (Schwein.:Fr.) (Anam. *Conoplea globosa* (Schwein.:Fr.) S. J. Hughes), cause of canker of oaks (*Quercus* spp.).

Comments: In *Plectania* Fuckel the apothecia are shallow cupulate to discoid and they sometimes have an orange to reddish exterior.

References: LeGal, 1958; Seaver, 1942; IM 14:C193.

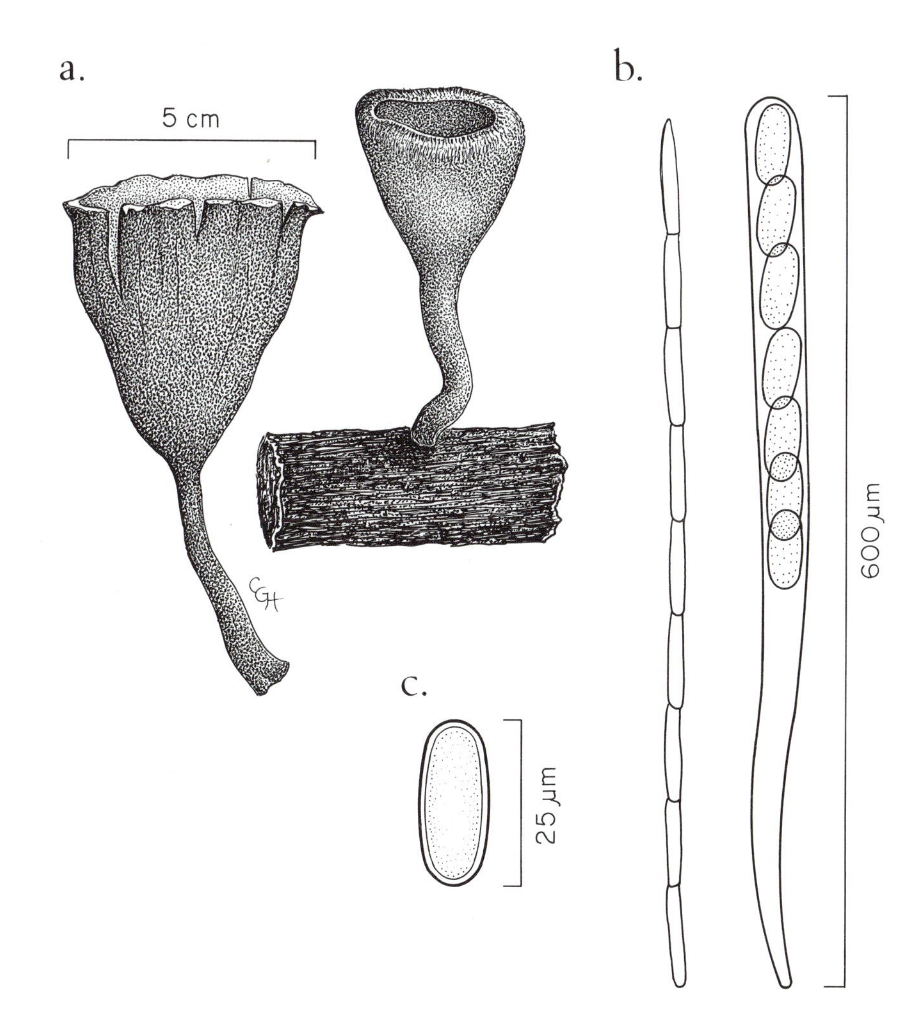

Urnula craterium. **a.** Left, mature apothecium; right, young apothecium on wood. **b.** Paraphysis and ascus with ascospores. **c.** Mature ascospore.

HYMENOSCYPHUS **S. F. Gray**

Ascoma an apothecium, up to 4 mm diameter, solitary or clustered; apothecium convex, flat or cupulate, stipitate, with stipe short or long, white, pink, yellow to brownish; ectal excipulum composed of parallel, thin-walled cells, medullary excipulum of loosely interwoven hyphae. Paraphyses cylindrical, sometimes slightly inflated at apex. Asci unitunicate, cylindric-clavate, pore blueing in iodine, 8-spored. Ascospores hyaline, 1-celled or occasionally 2-celled, elliptical to fusiform.

Anamorph: *Varicosporium*.

Habitat: On herbaceous plant debris.

Representative species: *Hymenoscyphus scutula* (Pers.:Fr.) W. Phillips.

Comments: Similar forms in which the ascus pore does not blue in iodine are placed in *Cudoniella* Sacc.

References: Dennis, 1963, 1981; Ellis and Ellis, 1985; Tubaki, 1966; BK 181-190; IM 30:C586.

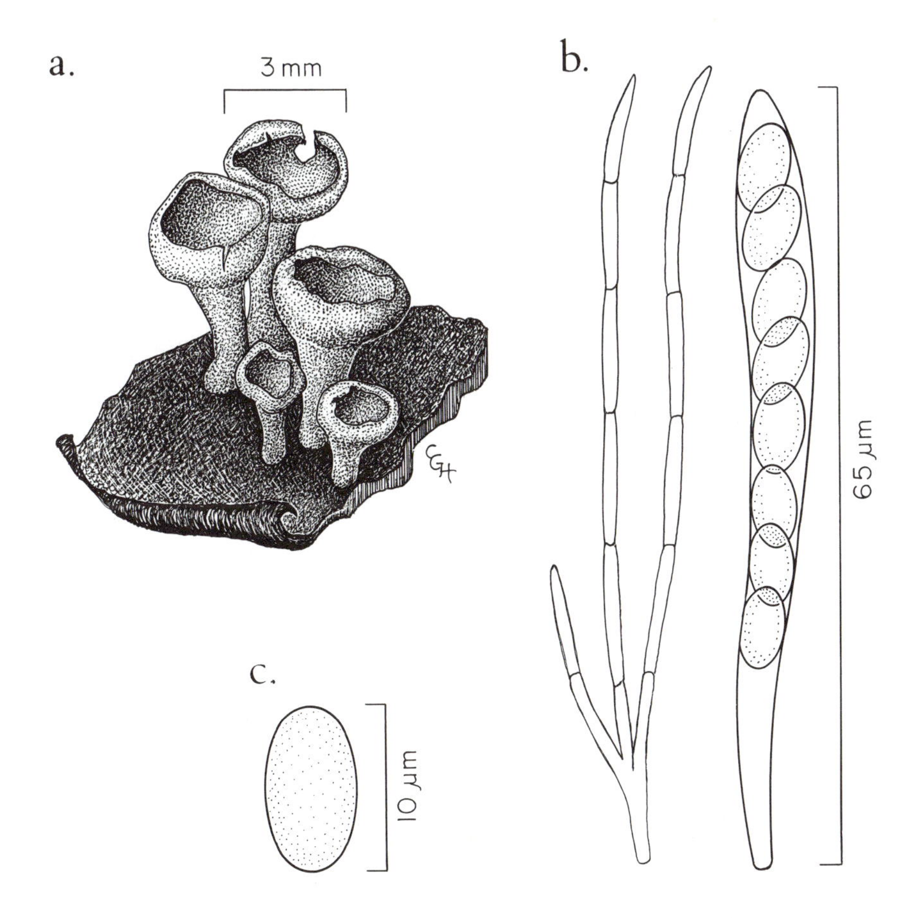

Hymenoscyphus robustior. **a.** Apothecia on dead wood. **b.** Paraphyses and ascus with ascospores. **c.** Mature ascospore.

CIBORIA **Fuckel**

Ascoma a stipitate apothecium, arising from a mummiform sclerotium formed in flowers or fruits of host that retain their original shape; apothecia cupulate to shallow saucer-shaped, often becoming flattened with age, up to 14 mm diam., usually brown, but rarely red, yellow or whitish; stipe slender, up to 30 mm long. Outermost cells of ectal excipulum globose, rarely covered by a layer of hyphae running parallel to surface; medullary excipulum of interwoven hyphae. Paraphyses filiform, sometimes branched, with slightly enlarged tips. Asci unitunicate, cylindric or subclavate, pore blueing in iodine, 4-8-spored; ascospores 1-celled, hyaline, ellipsoid to subellipsoid, sometimes inaequilateral, smooth or minutely roughened, sometimes with a thin gelatinous sheath. (Fig. 2E-G).

Anamorph: None reported, but spermatia formed.

Habitat: On flowers and fruits of flowering trees; the apothecia form on fallen, overwintered tissues.

Representative species: *Ciboria carunculoides* (Siegler & Jenk.) Whetzel, on ovaries of mulberry (*Morus* spp.).

Comments: In *Rutstroemia* P. Karst. the apothecia differ in being greenish in color.

References: Dennis, 1981; Ellis and Ellis, 1985; Schumacher, 1978; Whetzel and Buchwald, 1936; Whetzel and Wolf, 1945; Seaver, 1951; BK 146-148; FC 1; IM 44:C839.

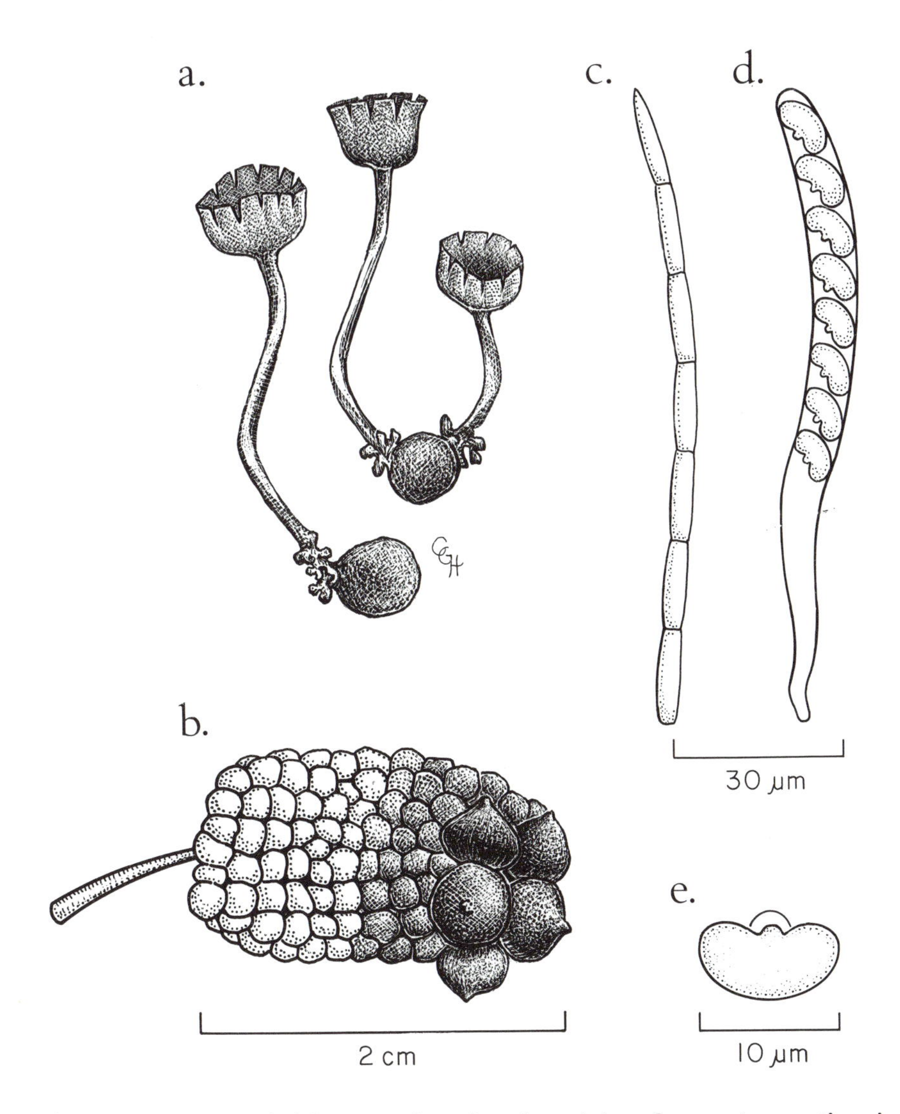

Ciboria carunculoides. **a.** Apothecia arising from stromatized ovaries. **b.** Mulberry with infected druplet at apex. **c.** Paraphysis. **d.** Ascus with ascospores. **e.** Mature ascospore.

MONILINIA Honey

Ascoma a stipitate apothecium, arising from mummified host tissues, usually fruits; apothecia cupulate to subdiscoid, up to 12 mm wide, yellow-brown to brown; stipe usually long. Medullary excipulum of textura intricata to textura porrecta; ectal excipulum of hyphae radiating outward, sometimes with swollen cells. Paraphyses filiform, slightly enlarged at apex. Asci unitunicate, cylindric to subclavate, apex blueing in iodine, 8-spored. Ascospores 1-celled, ellipsoid, sometimes dimorphic, hyaline or subhyaline, smooth, sometimes with a thin sheath.

Anamorph: *Monilia*; microconidia also formed.

Habitat: Parasitic on members of the Cornaceae, Ericaceae and Rosaceae.

Representative species: *Monilinia fructicola* (G. Wint.) Honey (Anam. *Monilia cinerea* var. *americana* Wormald), cause of brown rot of peach and other fruits.

Comments: In *Phaeosclerotinia* Hori the ascospores are brown.

References: Batra, 1983; Batra and Harada, 1986; Dennis, 1981; Ellis and Ellis, 1985; FC 38.

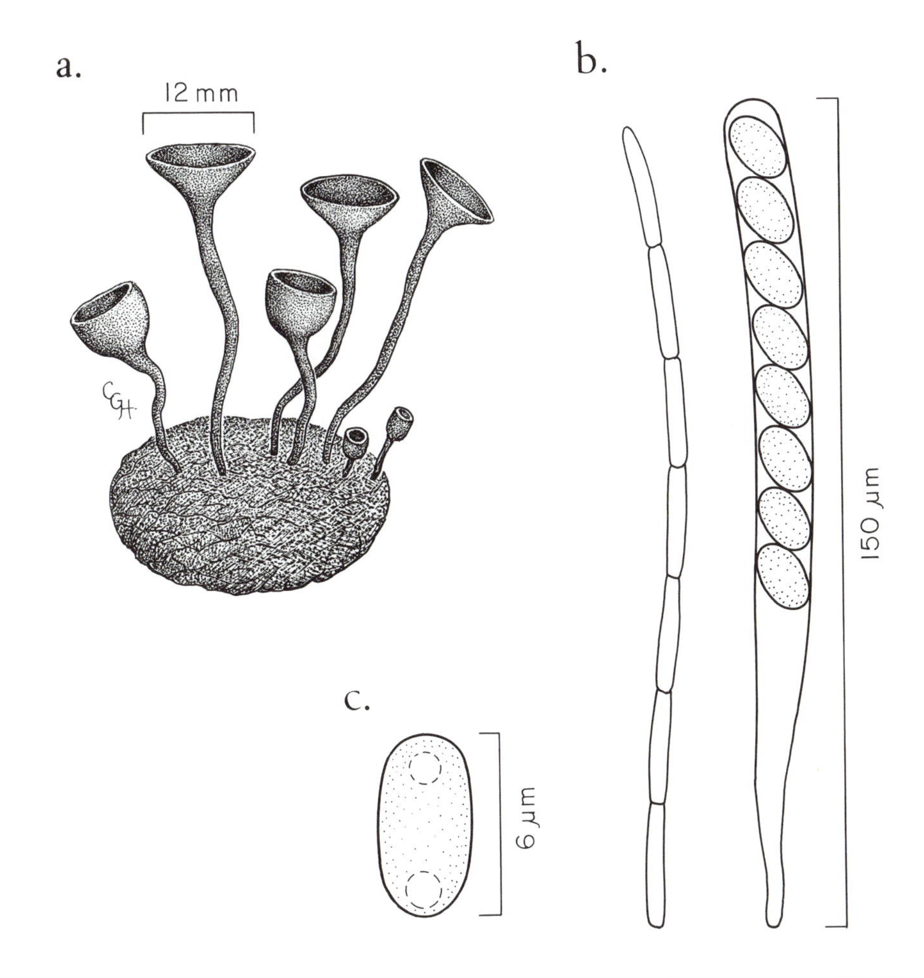

Monilinia fructicola. **a.** Apothecia arising from mummified peach fruit. **b.** Paraphysis and ascus with ascospores. **c.** Mature ascospore.

SCLEROTINIA **Fuckel**

Ascoma a stipitate apothecium, produced from a sclerotium that does not incorporate host tissues; apothecia 2-10 mm wide, cupulate to convex, cinnamon to umber in color, often darkening at margin. Subhymenium composed of textura intricata with light brown walls, usually gelatinous. Medullary excipulum composed of hyaline, loosely interwoven textura intricata. Ectal excipulum composed of textura prismatica, with outer cells inflated to globose, with hyaline to light brown walls. Asci interspersed with filiform, usually unbranched paraphyses. Asci unitunicate, cylindrical, tapering toward base, thin-walled but slightly thickened at apex, apical pore blueing in iodine, 8-spored. Ascospores hyaline, 1-celled, uniseriate in ascus, ellipsoid to somewhat flattened on one side, smooth, biguttulate.

Anamorph: None, but microconidia formed.

Habitat: Pathogenic on a wide range of herbaceous plants.

Representative species: *Sclerotinia sclerotiorum* (Lib.) de Bary, causing diseases of over 350 crops worldwide.

Comments: This genus was renamed *Whetzelinia* Korf & Dumont, but *Sclerotinia* was subsequently conserved nomenclaturally, making *Whetzelinia* a synonym of *Sclerotinia*. The genus *Ciborinia* Whetzel has been used for foliicolous species with small sclerotia.

References: Batra, 1960; Batra and Korf, 1959; Dennis, 1981; Domsch et al., 1980; Ellis and Ellis, 1985; Kohn, 1979; Seaver, 1951; BK 144-145; FC 90.

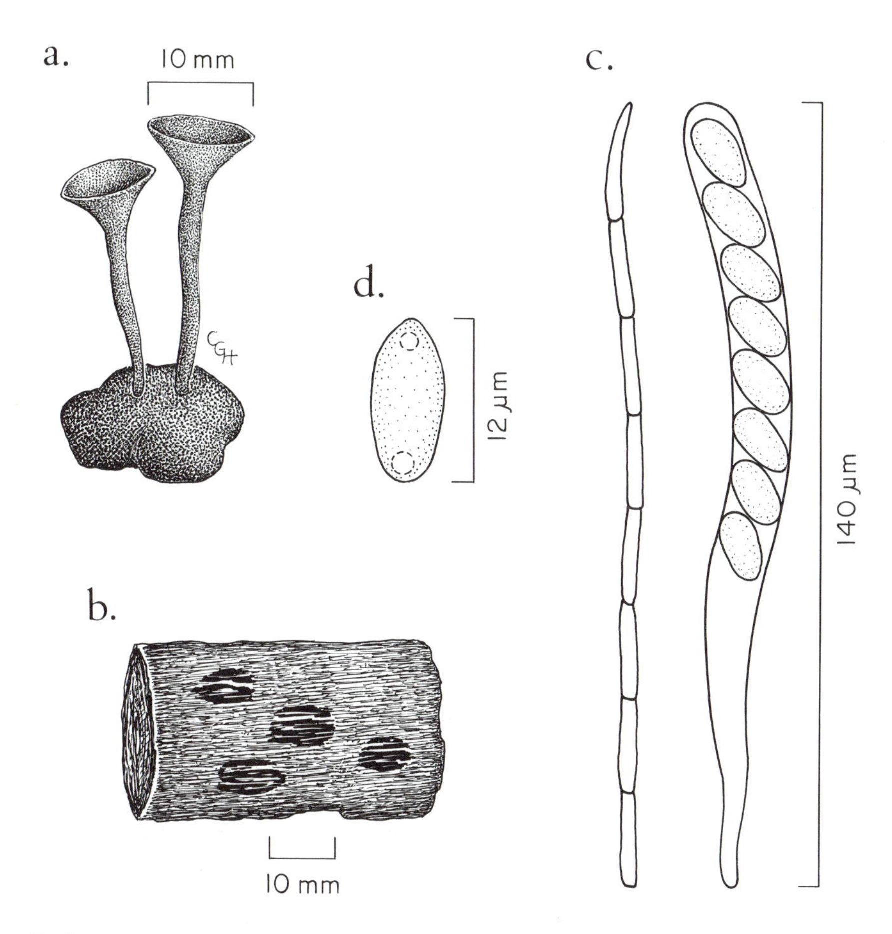

Sclerotinia sclerotiorum. **a.** Apothecia arising from sclerotium. **b.** Black sclerotia embedded in dead stem of dill. **c.** Paraphysis and ascus with ascospores. **d.** Mature ascospore.

MORCHELLA Dill. ex Pers.

Ascoma an erect, stipitate apothecium with a sponge-like pileus, up to 20 cm or more high, single, scattered; pileus subglobose, ovoid to narrowly conical, with anastomosing, irregular, sterile ridges separating shallow depressions bearing the hymenium, yellowish-brown to brown, often with ridges paler. Stipe hollow, brittle, subcyclindrical, sometimes bulbous at base, occasionally furrowed, yellowish or whitish-cream. Paraphyses septate near base, branched, with clavate apices. Asci unitunicate, operculate, subcylindrical, narrow at base, not blueing in iodine, 8-spored. Ascospores hyaline, 1-celled, broadly ellipsoidal to oblong-ellipsoidal, smooth, hyaline to subhyaline, pleurinucleate.

Anamorph: None reported.

Habitat: On soil in open areas.

Representative species: *Morchella esculenta* L., the common morel.

Comments: In *Verpa* Swartz the pileus is campanulate.

References: Dennis, 1981; Groves and Hoare, 1953; Rifai, 1968; Seaver, 1942; BK 1-6.

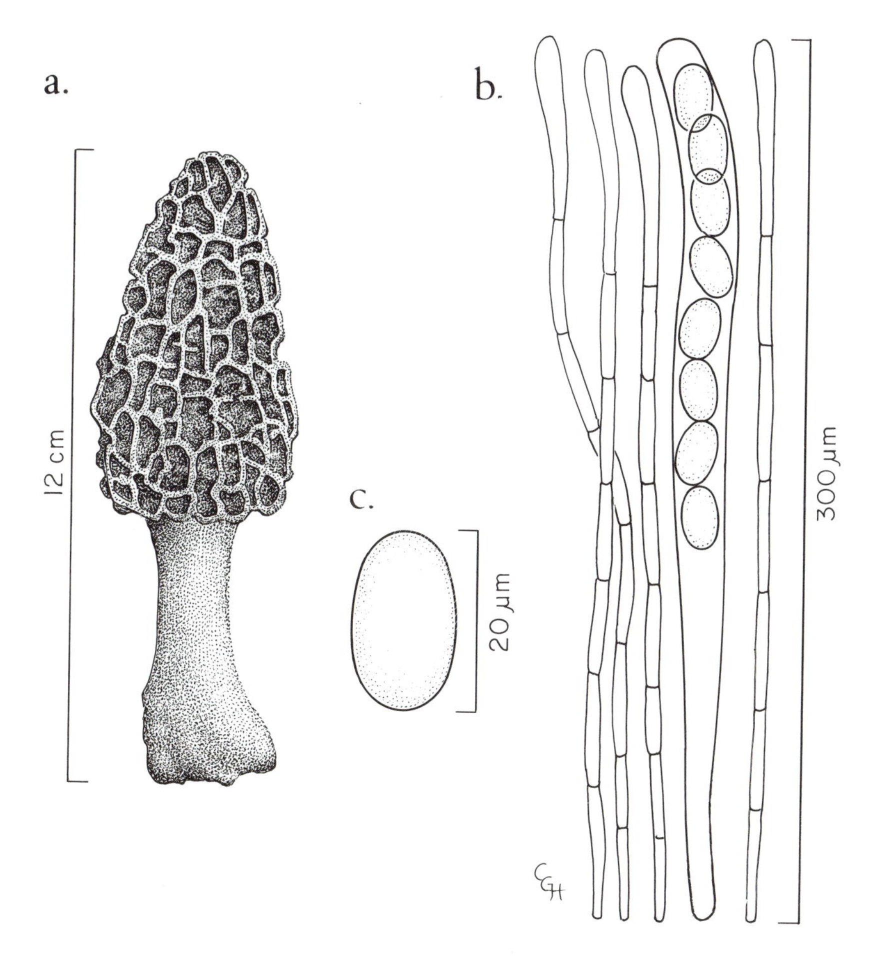

Morchella esculenta. **a.** Mature ascoma. **b.** Paraphyses and ascus with ascospores. **c.** Mature ascospore.

LEOTIA **Fr.**

Ascoma stipitate-capitate, caespitose, gregarious or scattered, consisting of a convex buff to greenish cap (up to 3 cm wide) borne on a terete greenish or yellow stipe (up to 9 cm high); the central core of stipe composed of interwoven hyphae with thick, gelatinous walls and surrounded by a cylinder of compact, non-gelatinous hyphae. The gelatinous tissues of the stipe expand outward to form the medullary tissues of the cap, on which the hymenium is formed. The surface of the stipe is covered by gelatinous hyphae similar to those of the core. Numerous filiform, branched paraphyses are interspersed with the asci in the hymenium. Asci unitunicate, clavate, with an apical pore, not blueing in iodine, 8-spored. Ascospores hyaline, 1-celled, sometimes becoming septate, subfusoid or narrowly ellipsoid, mostly multiguttulate, over 15 μm long.

Anamorph: None reported.

Habitat: On moist soil or dead wood.

Representative species: *Leotia lubrica* Fr., common on moist soil.

Comments: *Cudonia* Fr. and *Vibrissea* Fr. differ in having fleshy, nongelatinous ascomata; in *Cudonia* the ascospores are acicular or narrowly clavate, whereas in *Vibrissea* they are filiform.

References: Dennis, 1981; Mains, 1956; Seaver, 1951; BK 135-136.

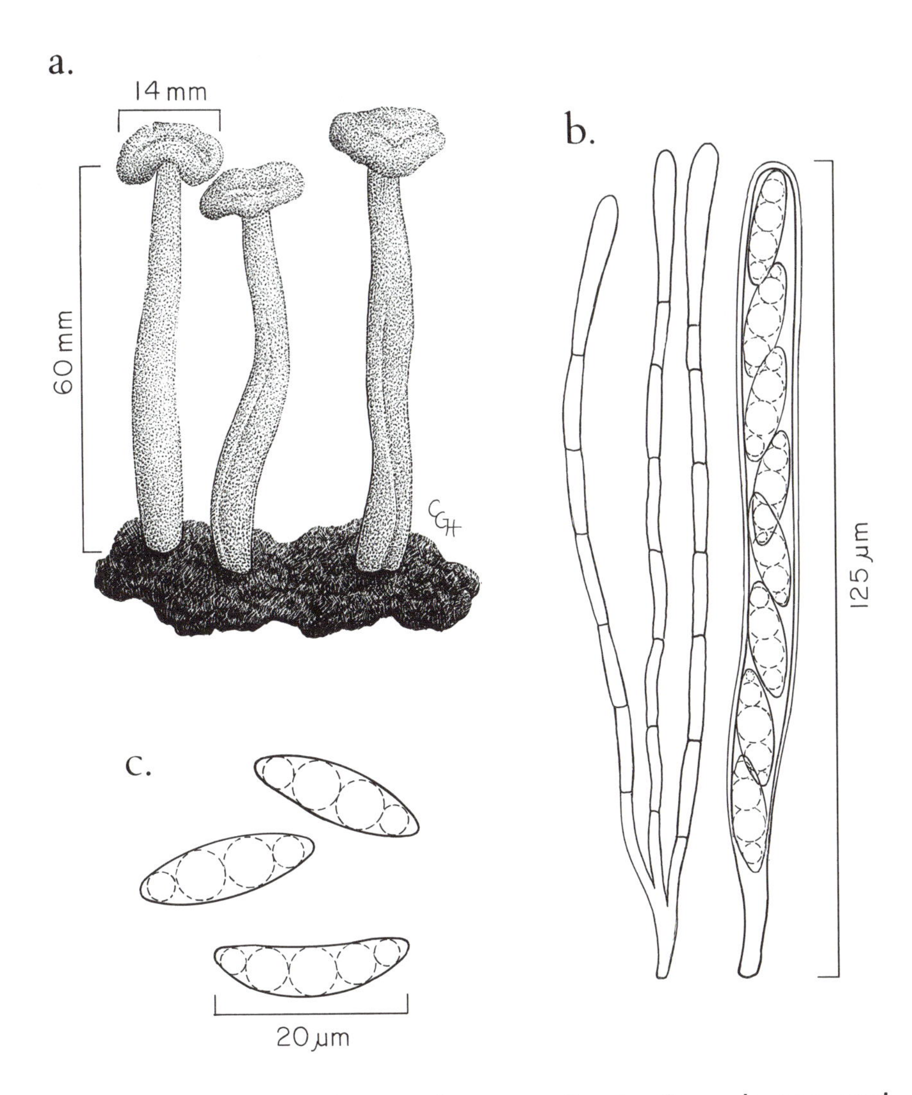

Leotia lubrica. **a.** Ascomata on soil. **b.** Paraphyses and ascus with ascospores. **c.** Mature ascospores.

HELVELLA L.

Ascoma an apothecium, solitary to gregarious, usually stipitate, but sometimes subsessile, cupulate, saddle-shaped or mitrate, up to 10 cm broad. Hymenial surface buff to brown, gray, black, or sometimes cream or white, smooth or rugose. Stipe variable, terete, compressed, or rounded and ribbed, of same colors as hymenium. Ectal excipulum textura angularis or a palisade of hyphal tips; medullary excipulum of textura intricata. Paraphyses simple, slender, with clavate tips. Asci unitunicate, operculate, cylindrical, nonamyloid, 8-spored. Ascospores hyaline, 1-celled, ellipsoid, oblong or fusiform, smooth to verruculose or verrucose-rugose, with a large central oil droplet, and often smaller droplets, tetranucleate at maturity.

Anamorph: None reported.

Habitat: On soil or occasionally rotting wood.

Representative species: *Helvella elastica* Fr., a common species in moist woods.

Comments: Cupulate species with fluted stipes have been placed in the genera *Acetabula* (Fr.) Fuckel or *Paxina* Kuntze.

References: Dennis, 1981; Kempton and Wells, 1970; Weber, 1972; BK 14-18(-27).

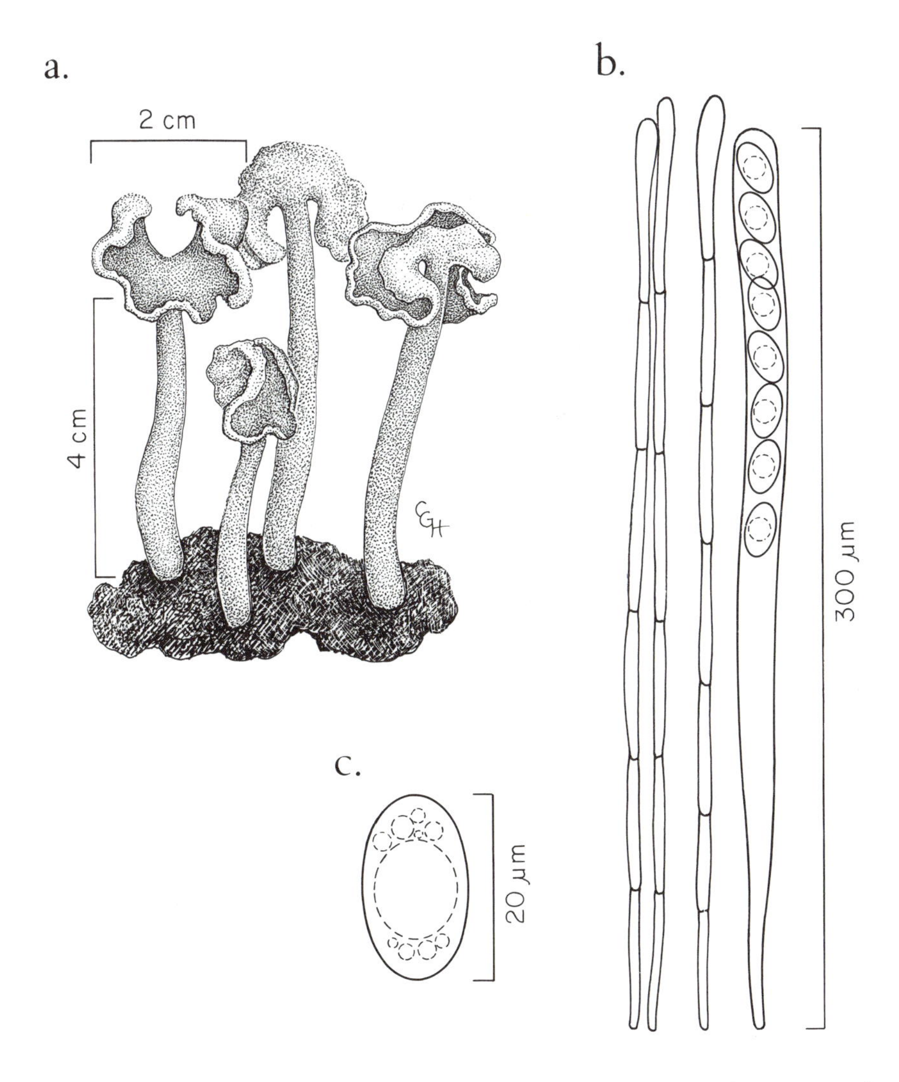

Helvella elastica. **a.** Apothecia on soil. **b.** Paraphyses and ascus with ascospores. **c.** Mature ascospore.

BLUMERIELLA **Arx**

Ascoma an apothecium, immersed in host tissues, overlying tissues erumpent but apothecium remaining depressed; apothecium flat to shallow-convex; ectal excipulum of brown-walled textura globulosa or textura angularis. Paraphyses filiform, septate, somewhat inflated at apex. Asci unitunicate, clavate to subclavate, with pore blueing in iodine. Ascospores 1-celled, but sometimes with a median septum, hyaline, ellipsoid to elongate-fusiform, narrower in middle.

Anamorph: *Phloeosporella* (*Cylindrosporium*).

Habitat: Parasitic on living leaves of rosaceous plants, but apothecia may develop in dead, overwintered leaves.

Representative species: *Blumeriella jaapii* (Rehm) Arx [Anam. *Phloeosporella padi* (Lib.) Arx], cause of shot-hole disease of cherry *(Prunus* spp.). This species was formerly classified as *Coccomyces hiemalis* Higgins and *Higginsia hiemalis* (Higgins) Nannf.

Comments: This genus differs from *Pseudopeziza* Fuckel in the presence of a well-developed excipulum at the sides of the hymenium.

References: Arx, 1961; Ellis and Ellis, 1985; Williamson and Bernard, 1988.

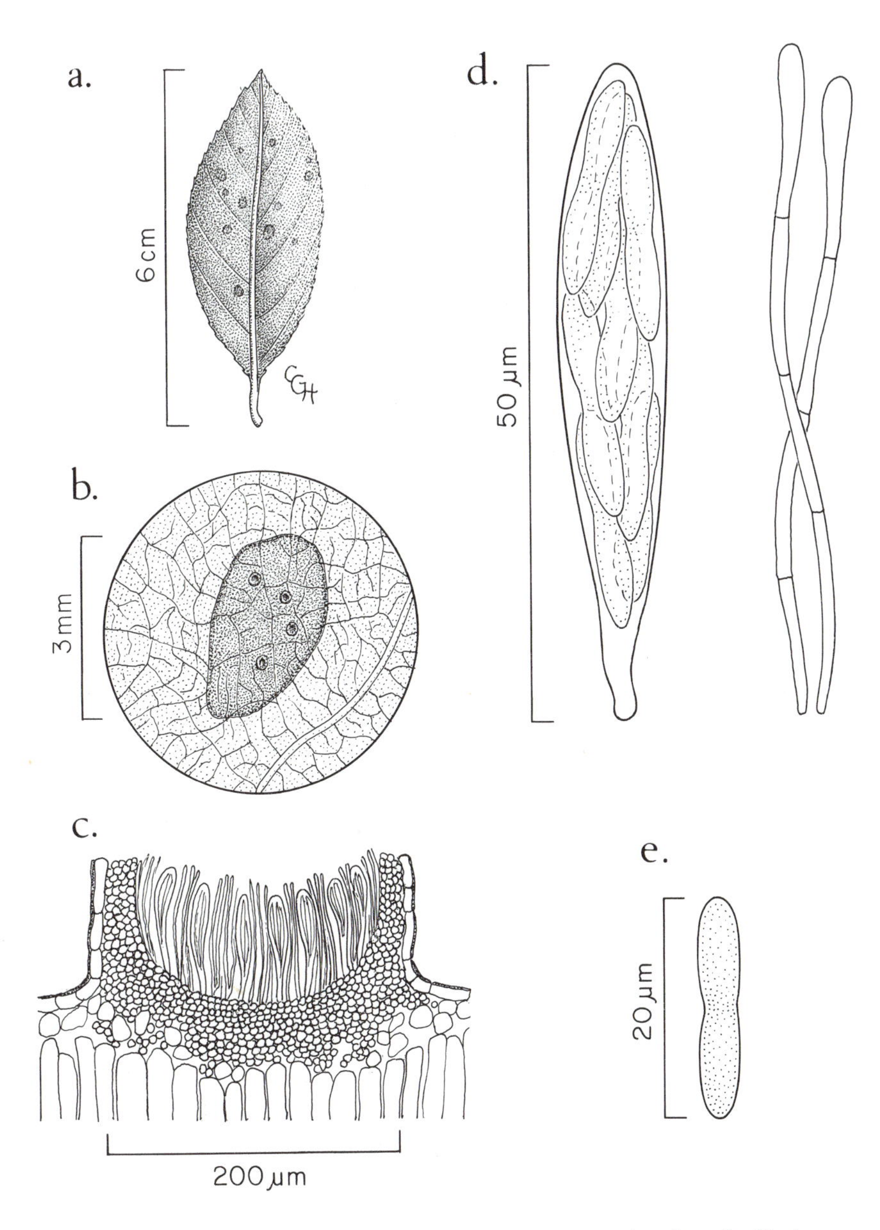

Blumeriella haddenii. **a.** Lesions on leaf of *Spiraea.* **b.** Close-up of lesion with apothecia. **c.** Section through apothecium with asci and paraphyses. **d.** Paraphyses and ascus with ascospores. **e.** Mature ascospore.

PSEUDOPEZIZA **Fuckel**

Ascoma an apothecium immersed in host tissue, single, erumpent at maturity; apothecium disc-shaped, up to 0.5 mm diameter, brown to yellowish, with soft flesh, lacking a pseudoparenchymatous ectal excipulum. Paraphyses filamentous. Asci unitunicate, pore not blueing in iodine, cylindric-clavate, 8-spored. Ascospores 1-celled, or occasionally 2-celled, hyaline, ovoid.

Anamorph: None reported.

Habitat: On living leaves of herbaceous plants.

Representative species: *Pseudopeziza trifolii* (Biv.-Bern.:Fr.) Fuckel, cause of leafspot of clover *(Trifolium* spp.); *P. medicaginis* (Lib.)Sacc. on alfalfa *(Medicago* spp.) is regarded as a forma specialis of *P. trifolii.*

Comments: This genus differs from *Blumeriella* Arx in the lack of a pseudoparenchymatous excipulum at the sides of the hymenium. *Leptotrochila* P. Karst. differs in that the apothecium arises from a stroma.

References: Ellis and Ellis, 1985; Schuepp, 1959; BK 284; CMI 636, 637; FC 229.

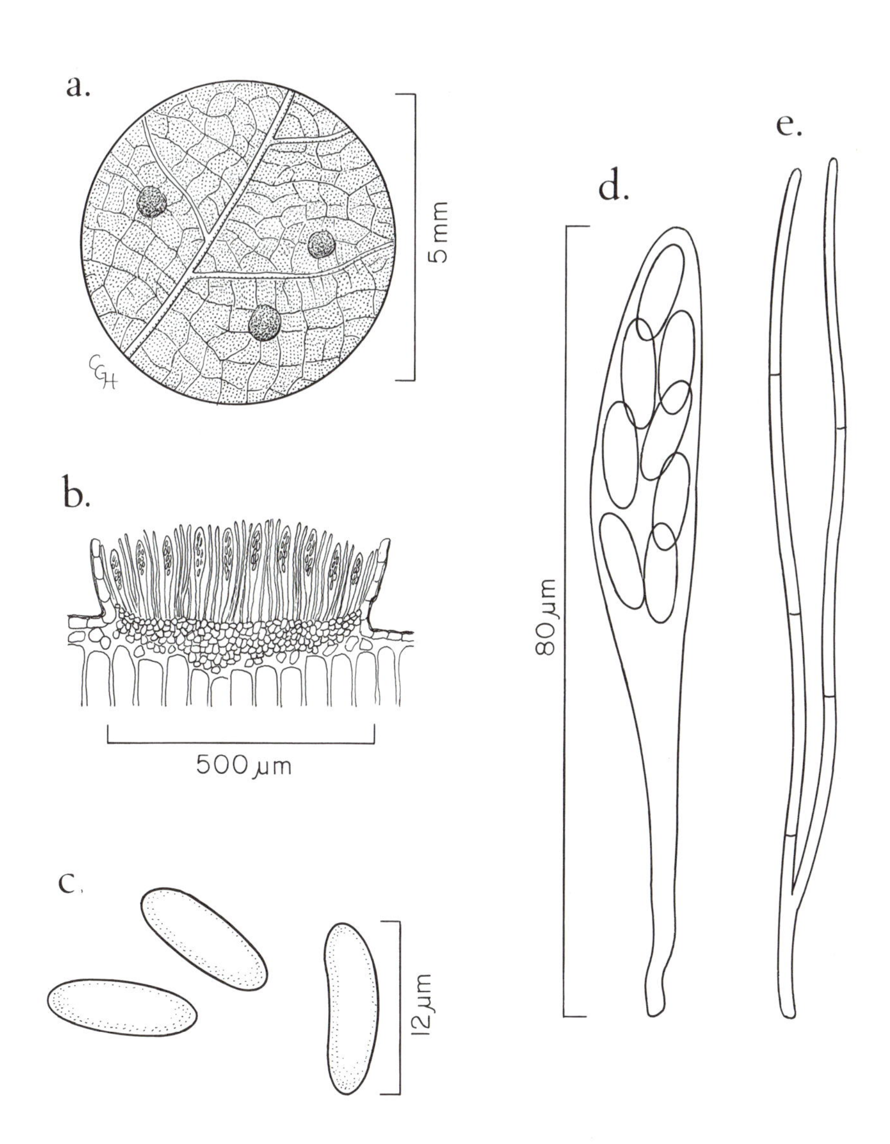

Pseudopeziza trifolii f. sp. *medicaginis.* **a.** Erumpent apothecia on alfalfa leaf. **b**. Section through mature apothecium. **c**. Mature ascospores. **d**. Ascus with ascospores. **e**. Paraphyses.

PLOIODERMA Darker

Ascoma an elliptical apothecium, shiny black, arising from a subepidermal stroma, erumpent above host epidermis and opening by a slit at maturity. Paraphyses filamentous, septate, abundant. Asci unitunicate, clavate to cylindrical, pore not blueing in iodine, 8-spored. Ascospores hyaline, fusoid to clavate, often slightly narrower in middle, 1-celled, but becoming septate upon germination, surrounded by a gelatinous sheath.

Anamorph: *Leptostroma.*

Habitat: On living needles of conifers.

Representative species: *Ploioderma lethale* (Dearn.) Darker, cause of needle cast disease of conifers.

Comments: *Virgella* Darker, *Hypoderma* De Not., and *Meloderma* Darker also have rod-like ascospores. In *Virgella* the ascomata are linear, whereas in *Hypoderma* and *Meloderma* they are elliptical; in *Hypoderma* the asci are clavate, and in *Meloderma* they are cylindrical.

References: Czabator, 1976; Darker, 1932, 1967; Hunt and Ziller, 1978; CMI 570, 799; IM 21:C304.

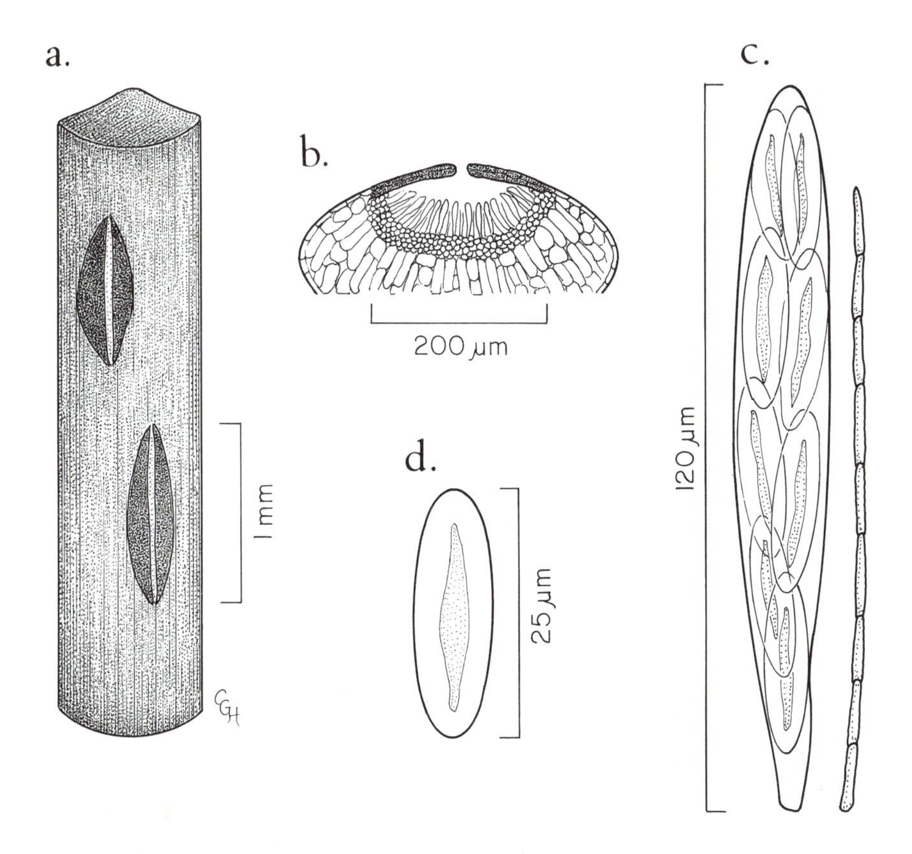

Ploioderma lethale. **a.** Two hysterothecia on pine needle. **b.** Section through hysterothecium on pine needle. **c.** Paraphysis and ascus with ascospores. **d.** Mature ascospore with gelatinous sheath.

PHACIDIUM Fr.

Ascoma apothecioid, circular, immersed in host tissues, subcuticular, subepidermal or subhypodermal, erupting at maturity, splitting open to expose the hymenium. Ascoma flattened to concave, upper covering composed of dark brown pseudoparenchyma cells, with an interior layer of pale brown cells. Basal stroma present or absent; if present, it is pseudoparenchymatous. Subhymenium composed of angular or globose cells. Paraphyses simple or branched, septate, often anastomosing, invested in mucilage. Asci unitunicate, club-shaped, with an amyloid apical ring, 4-8-spored. Ascospores 1-celled, ellipsoid to fusiform, hyaline, smooth.

Anamorphs: *Apostrasseria* and *Ceuthospora.*

Habitat: Parasitic or saprobic on stems and leaves of vascular plants.

Representative species: *Phacidium abietis* (Dearn.) J. Reid & Cain, cause of snow-blight disease of fir (*Abies* and *Pseudotsuga*).

Comments: In *Lophophacidium* Lagerberg the ascomata are elliptical and straw-colored, and they open by a longitudinal slit. *Phacidiostroma* Höhn. has been used for species which occupy the entire leaf thickness, forming amphigenous apothecia; it is regarded as a synonym of *Phacidium.*

References: Dicosmo et al., 1984; CMI 652, 653; IM 25:C433.

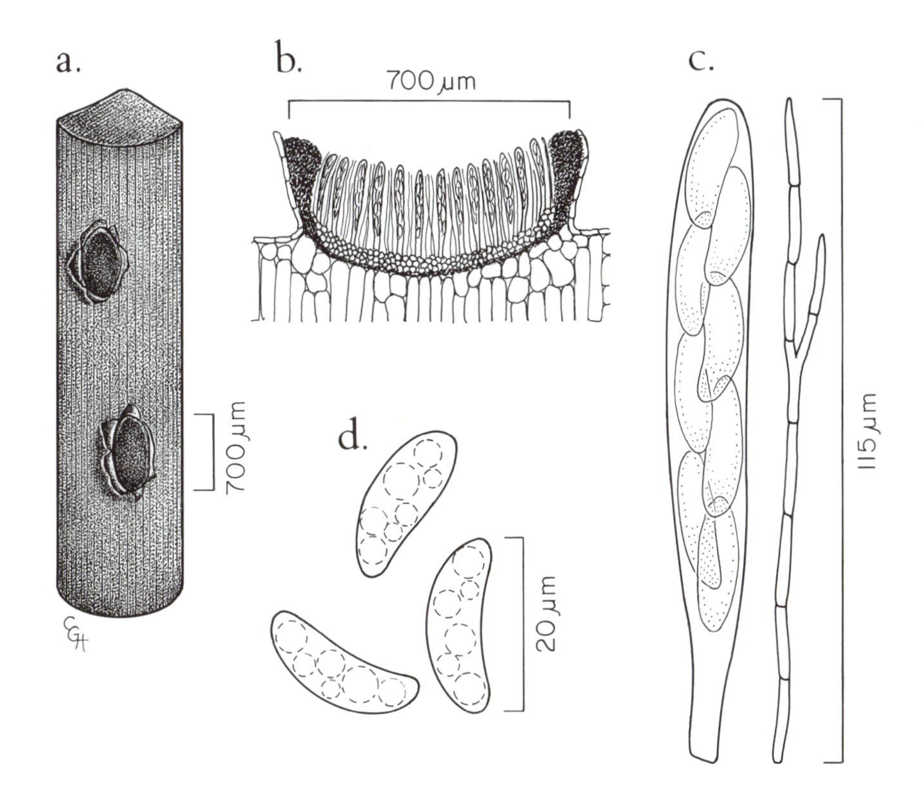

Phacidium infestans. **a.** Erumpent apothecia on pine needle. **b**. Section through apothecium on pine needle. **c**. Paraphysis and ascus with ascospores. **d**. Mature ascospores.

DIATRYPE Fr.

Ascoma an ostiolate perithecium, immersed in a stroma; stroma formed entirely of fungus tissue, erumpent through bark, discoid to pulvinate or hemisphaerical, discrete or wide-spreading, with surface flat or slightly convex. Perithecia usually arranged in a single layer, with ostiolar necks emerging straight to surface. Ostioles 3-4-sulcate, flattened and circular or subconical. Asci unitunicate, spindle-shaped, long-stipitate, with a refractive, simple amyloid ring at apex, small, 8-spored. Ascospores 1-celled, allantoid, hyaline but often greenish to brownish.

Anamorph: Coelomycetes with sympodial or annelidic modes of conidiogenesis not referable to current genera with certainty.

Habitat: On woody substrates.

Representative species: *Diatrype virescens* (Schwein.) M. A. Curtis on dead branches of beech *(Fagus* spp.).

Comments: In *Diatrypella* (Ces. & De Not.) Sacc. the asci are polysporous.

References: Dennis, 1981; Ellis and Ellis, 1985; Ellis and Everhart, 1892; Glawe and Rogers, 1984; Viégas, 1944; BK 356-358; FC 72, 73.

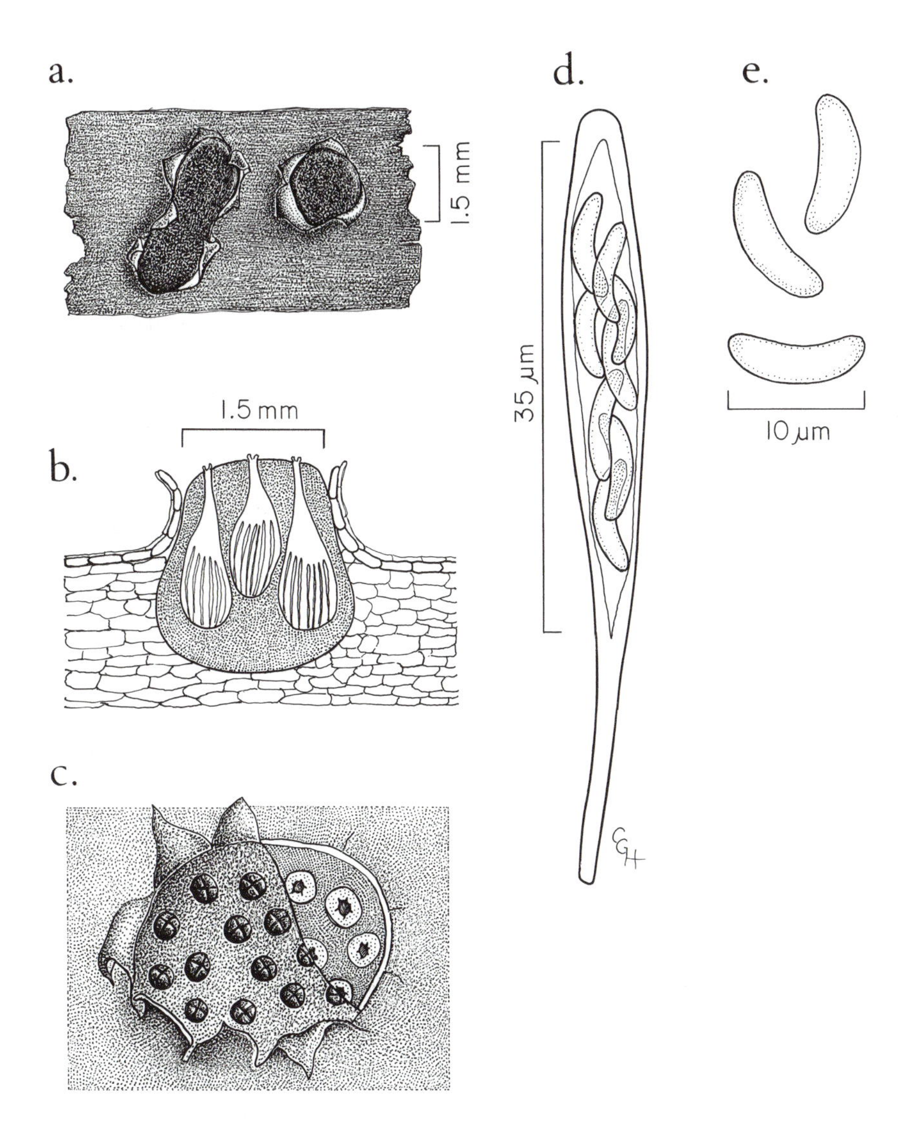

Diatrype virescens. **a.** Discs of stromata erumpent through host periderm. **b**. Section through stroma containing three perithecia. **c**. Surface of disc with sulcate ostioles; right portion cut away to show ostiolar necks in stroma. **d**. Ascus with ascospores. **e**. Mature ascospores.

EUTYPA Tul. & C. Tul.

Ascoma an ostiolate perithecium, immersed in a stroma; stroma usually wide-spreading, eutypoid, embedded in substrate, often with poorly defined margins, but sometimes discrete, incorporating host tissues, surface black, often roughened. Perithecia forming a single layer in stroma, subglobose, with ostiolar necks separately erumpent, separated by green or white pseudoparenchyma tissue or by blackened or nearly unaltered host tissues. Ostioles subconical, 3-5-sulcate or asulcate. Asci unitunicate, spindle-shaped, long-stipitate, with slightly amyloid apical ring, 8-spored. Ascospores 1-celled, allantoid to suballantoid, subhyaline to dark brown.

Anamorphs: Coelomycetes with sympodial and annelidic modes of conidiogenesis not referable to current genera with certainty.

Habitat: In decorticated wood or occasionally in bark, or parasitic in woody plants.

Representative species: *Eutypa armeniacae* Hansf. & M. V. Carter, cause of canker of apricot.

Comments: *Cryptosphaeria* Grev. occurs mainly in bark and has 8-spored asci; in *Cryptovalsa* Ces. & De Not. ex Fuckel the asci are polysporous.

References: Dennis, 1981; Ellis and Ellis, 1985; Ellis and Everhart, 1892; Glawe and Rogers, 1984; Rappaz, 1987; Viégas, 1944; BK 353-355; CMI 436.

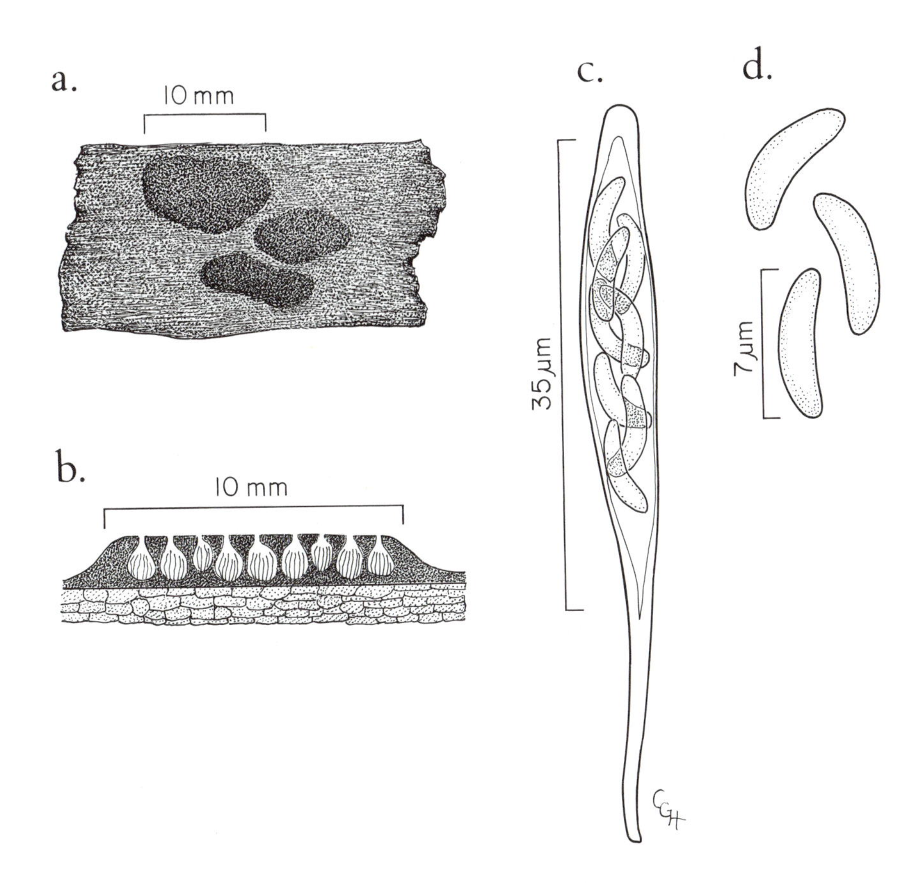

Eutypa armeniacae. **a.** Stromata on wood. **b**. Section through stroma with single row of perithecia. **c**. Ascus with ascospores. **d**. Mature ascospores.

EUTYPELLA (Nitschke) Sacc.

Ascoma an ostiolate perithecium, immersed in a stroma composed of both host and fungal tissue; stroma valsoid, circular to elliptical, delimited in host by a blackened zone, with a black, erumpent ostiolar disc. Perithecia polystichous in stroma, with elongate ostiolar necks that converge; ostioles indistinctly 3-5-sulcate. Asci unitunicate, clavate to spindle-shaped, long-stipitate, with nonamyloid apical rings, 8-spored. Ascospores 1-celled, allantoid, subhyaline to subolivaceous.

Anamorphs: Coelomycetes with sympodial and annellidic modes of conidiogenesis not referable to current genera with certainty.

Habitat: Occurring on dead branches of hardwoods.

Representative species: *Eutypella scoparia* (Schwein.:Fr.) Ellis & Everh., on woody tissues of various plants.

Comments: In *Quaternaria* Tul. & C. Tul. the ascomata are sparse and weakly erumpent, and the stroma is often much reduced.

References: Dennis, 1981; Ellis and Ellis, 1985; Glawe and Rogers, 1984; Munk, 1957; Rappaz, 1987; Viégas, 1944; BK 361; IM 1:C12.

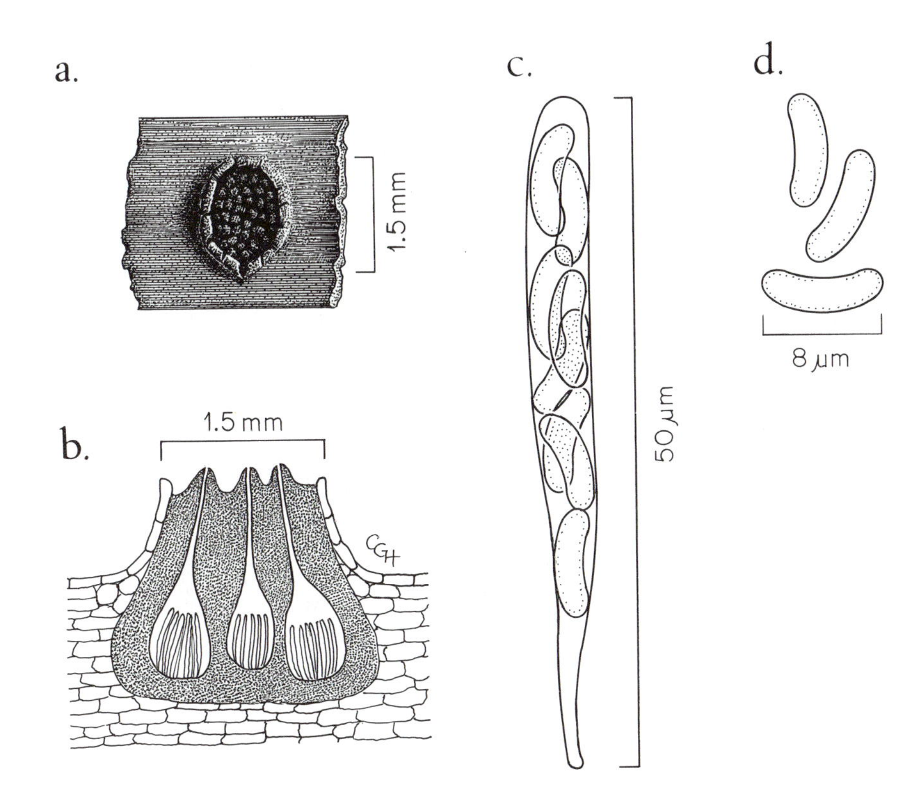

Eutypella citricola. **a.** Stroma erumpent through periderm of host. **b**. Section through stroma with perithecia. **c**. Ascus with ascospores. **d**. Mature ascospores.

VALSA Fr.

Ascoma an ostiolate perithecium, formed in a prosenchymatous valsoid stroma; stroma immersed in host tissues, black, erumpent. Perithecia clustered or in a circle in stroma, subglobose, with long ostiolar necks converging through disc. Ascomal wall dark, outer cells isodiametric and thick-walled, inner cells hyaline, thin-walled and flattened. Asci unitunicate, cylindric to ellipsoid or clavate, with a nonamyloid refractive apical ring, lying free in perithecium at maturity, 4-8-spored. Ascospores hyaline, 1-celled, allantoid to subcylindric, less than 30 µm long.

Anamorph: *Cytospora*.

Habitat: In branches of conifers and deciduous trees and shrubs.

Representative species: *Valsa ambiens* (Pers.: Fr.) Fr. [Anam. *Cytospora leucosperma* (Pers.:Fr) Fr.], on numerous hardwoods.

Comments: In *Ophiovalsa* Petr. the ascospores are 1-several-celled and usually more than 30 µm long.

References: Barr, 1978; Dennis, 1981; Ellis and Ellis, 1985; Ellis and Everhart, 1892; Kobayashi, 1970; Munk, 1957; Spielman, 1985; Viégas, 1944.

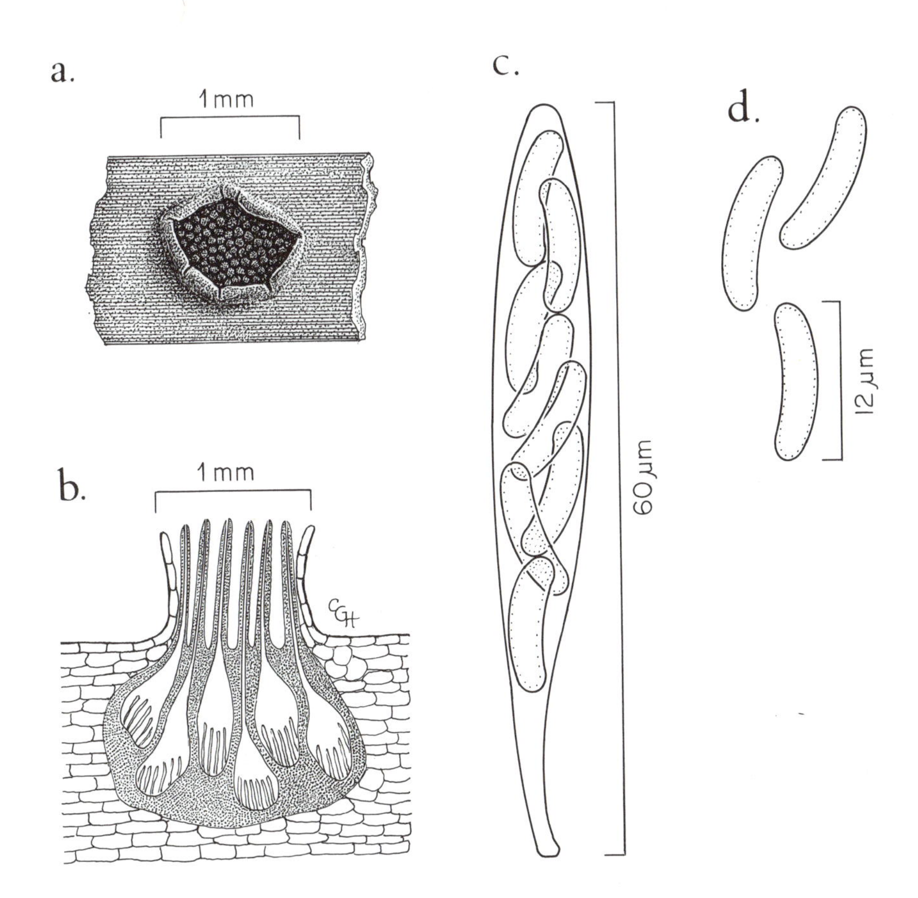

Valsa ambiens. **a.** Stroma erumpent through host periderm. **b.** Section through stroma with perithecia. **c.** Ascus with ascospores. **d.** Mature ascospores.

ENDOTHIA Fr.

Ascoma an ostiolate perithecium, immersed in a diatrypoid, pseudoparenchymatous stroma; stroma immersed in host tissues, brightly colored, erumpent through host periderm at maturity. Perithecia in one or several layers, subglobose, upright, with ostiolar necks erumpent separately. Ascomal wall dark brown, outer cells flattened and thick-walled, inner cells hyaline and thin-walled. Asci unitunicate, clavate to ellipsoid, with nonamyloid refractive ring in apex, often with deliquescent base, and lying free in perithecium at maturity, 8-spored. Ascospores hyaline, 1-celled, allantoid to cylindric.

Anamorph: *Endothiella.*

Habitat: On woody hardwood stems, especially oaks *(Quercus* spp.).

Representative species: *Endothia gyrosa* (Schwein.:Fr.) Fr. (Anam. *Endothiella gyrosa* Sacc.).

Comments: The species of *Endothia* with 2-celled ascospores and a valsoid stroma are placed in *Cryphonectria* (Sacc.) Sacc. *Valseutypella* Höhn. differs in having the stroma composed of thick-walled cells, a dark exterior, and a light interior.

References: Barr, 1978; Ellis and Everhart, 1892; Kobayashi, 1970; Micales and Stipes, 1987; Roane et al., 1986; CMI 449.

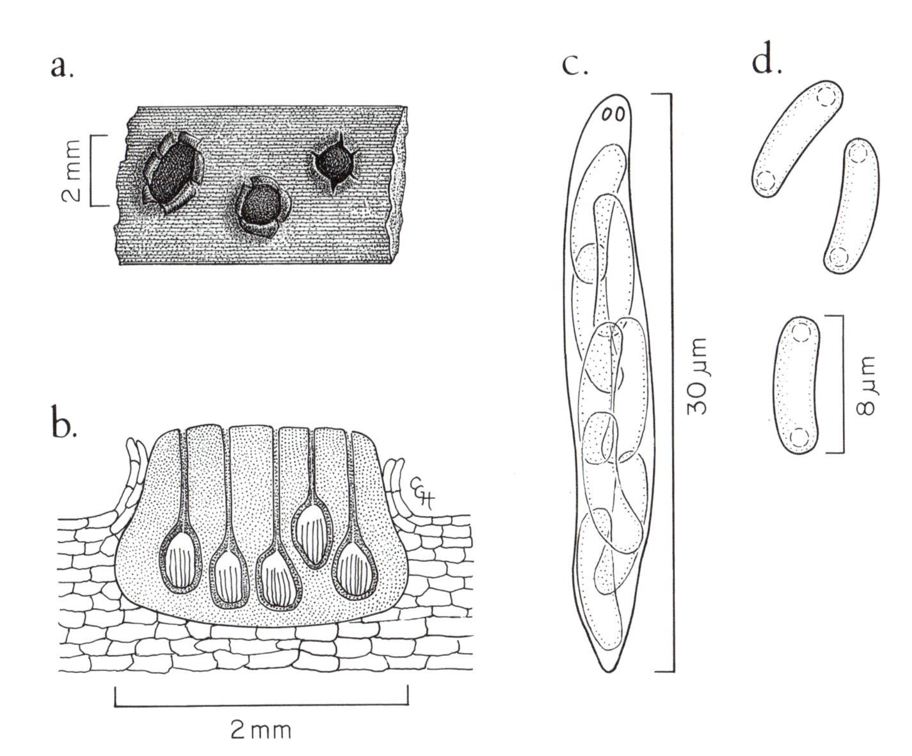

Endothia gyrosa. **a.** Perithecial stromata erumpent through periderm of host. **b**. Section through stroma with perithecia. **c**. Ascus with ascospores. **d**. Mature ascospores.

MICROASCUS **Zukal**

Ascoma perithecioid, superficial or partially immersed in substrate, spherical or obpyriform, ostiolate, with ostiolar papilla or cylindrical neck, often with hairs or setae, especially around ostiole, brown or black. Ascomal wall composed of isodiametric or slightly flattened pseudoparenchyma cells, walls of outer cells pigmented and slightly thickened. Asci catenulate or in a fascicle, obovoid or broadly clavate, 8-spored, with walls evanescent at maturity. Ascospores 1-celled, asymmetrical, reniform, heart-shaped or triangular, smooth, dextrinoid when young, yellowish to straw color at maturity, with a single basal germ pore. (Fig. 5D-E).

Anamorphs: *Scopulariopsis* and *Wardomyces.*

Habitat: In soil, dung, seeds and plant debris.

Representative species: *Microascus trigonosporus* C. W. Emmons & B. O. Dodge.

Comments: Species with narrow ascospores that lack germ pores and are produced in ostiolate or nonostiolate ascomata that lack an anamorph have been placed in *Pithoascus* Arx.

References: Arx, 1975b; Barron et al., 1961; Domsch et al., 1980; Matsushima, 1975; Morton and Smith, 1963; FC 180; IM 8:C108

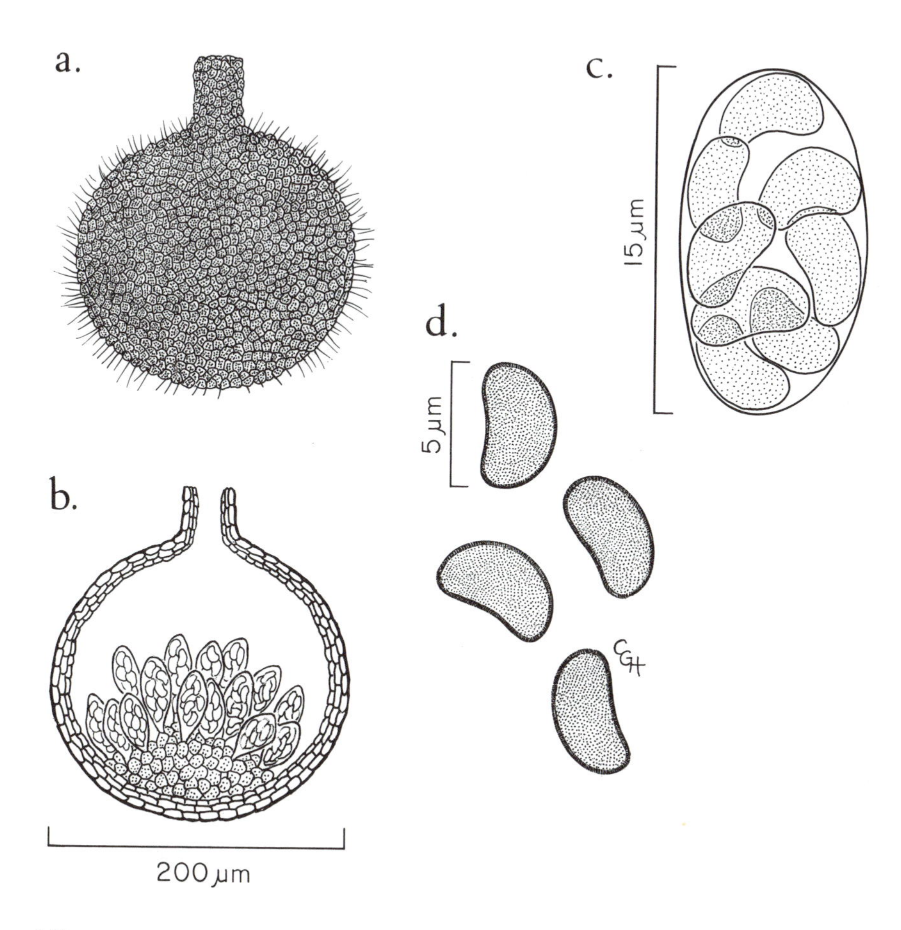

Microascus cirrosus. **a.** Exterior of ascoma. **b**. Section through ascoma with asci. **c**. Ascus with ascospores. **d**. Mature ascospores.

MELANOSPORA Corda

Ascoma an ostiolate or nonostiolate perithecium, superficial, solitary to gregarious, globose to subglobose, glabrous to tomentose, ostiolate species with short or long ostiolar neck that is ringed by hyaline setae. Ascomal wall composed of polyhedral pseudoparenchymata cells, membraneous and translucent, usually thin, pale yellow to reddish brown. Centrum composed of pseudoparenchyma cells. Asci clavate, undifferentiated at apex, evanescent at maturity, 4-8-spored. Ascospores 1-celled, ellipsoidal to citriform, occasionally discoid or fusiform, with a depressed germ pore at each end, brown, smooth.

Anamorphs: *Acremonium, Chlamydomyces, Harzia, Paecilomyces,* and *Proteophiala.*

Habitat: Parasitic on fungi or saprobic on plant debris; in seeds; often occurs in cultures in association with other fungi.

Representative species: *Melanospora zamiae* Corda, on a wide range of substrates.

Comments: Traditionally *Melanospora* has included species with diverse types of ascospores, but Cannon and Hawksworth restrict the genus to species that have ascospores with smooth walls and a depressed pore at each end. Species that have coarsely reticulate ascospores with umbonate pores are placed in *Sphaerodes* Clements, whereas species with indistinct ornamentation and slightly apiculate ascospores are placed in *Persiciospora* P. Cannon & D. Hawksworth. Species with smooth-walled, cylindrical ascospores with depressed apical pores are placed in *Syspastospora* P. Cannon & D. Hawksworth, whereas those species that have ascospores with smooth walls, a germ slit and are cuboid-ellipsoid in shape are accommodated in *Scopinella* Lév. *Sphaeroderma* Fuckel has been used for species

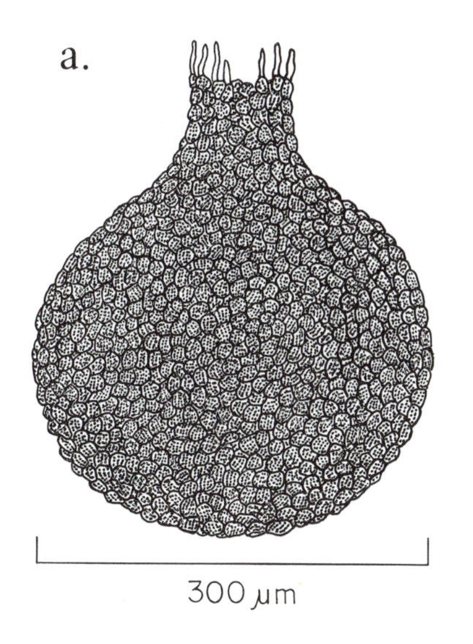

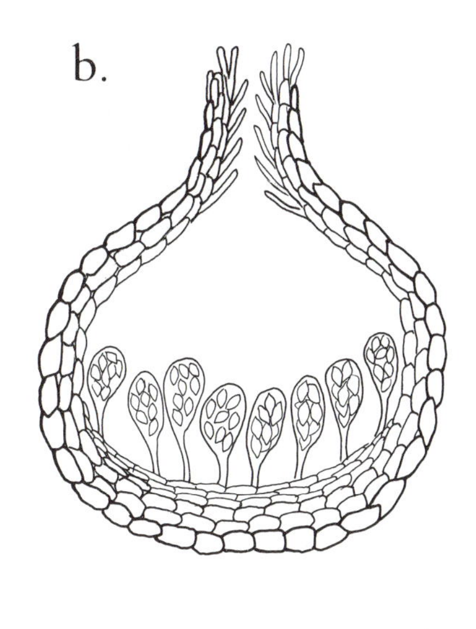

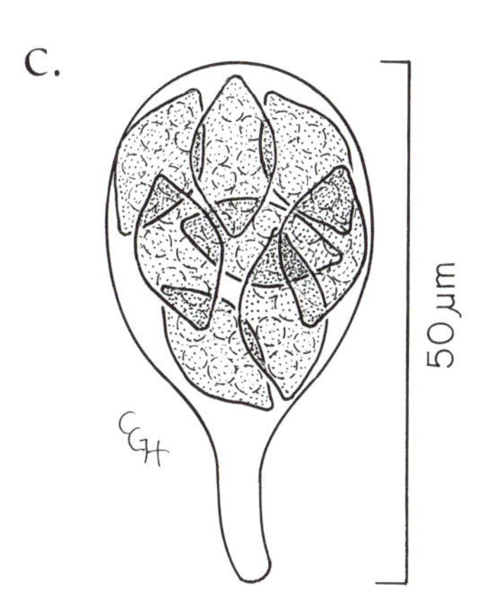

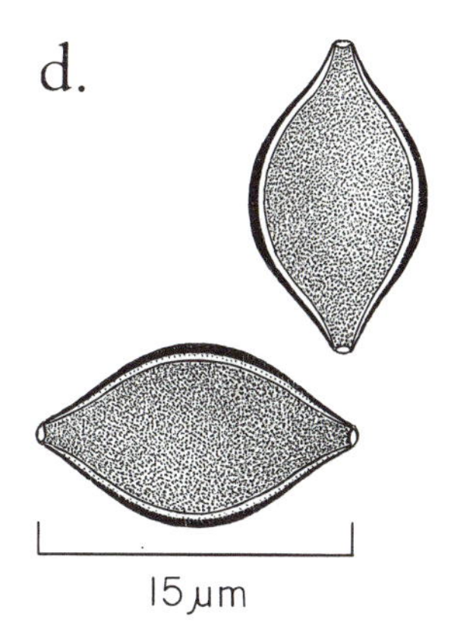

Melanospora zamiae. **a.** Exterior of perithecium. **b**. Section through perithecium with asci. **c**. Ascus with ascospores. **d**. Mature ascospores.

that lack an ostiolar neck and *Microthecium* Corda has been used for species with nonostiolate ascomata; both are regarded as synonyms of *Melanospora.*

References: Cannon and Hawksworth, 1982; Doguet, 1955; Ellis and Ellis, 1985; Ellis and Everhart, 1892; Hawksworth and Udagawa, 1977; Matsushima, 1975; FC 82, 83; IM 3:C32.

→

Fig. 3. **A**. Ascoma of *Ceratocystis fimbriata*. X80. **B-C**. *Glomerella cingulata*. **B**. Perithecium. X284. **C**. Ascus with ascospores. X956. **D**. Mature ascospore of *C. fimbriata*. X1165. **E**. Mature ascospore of *G. cingulata*. X1250. **F-H**. *Gelasinospora retispora*. **F**. Perithecium. X102. **G**. Ascus with ascospores. X400. **H**. Ascospores with pitted walls. X600.

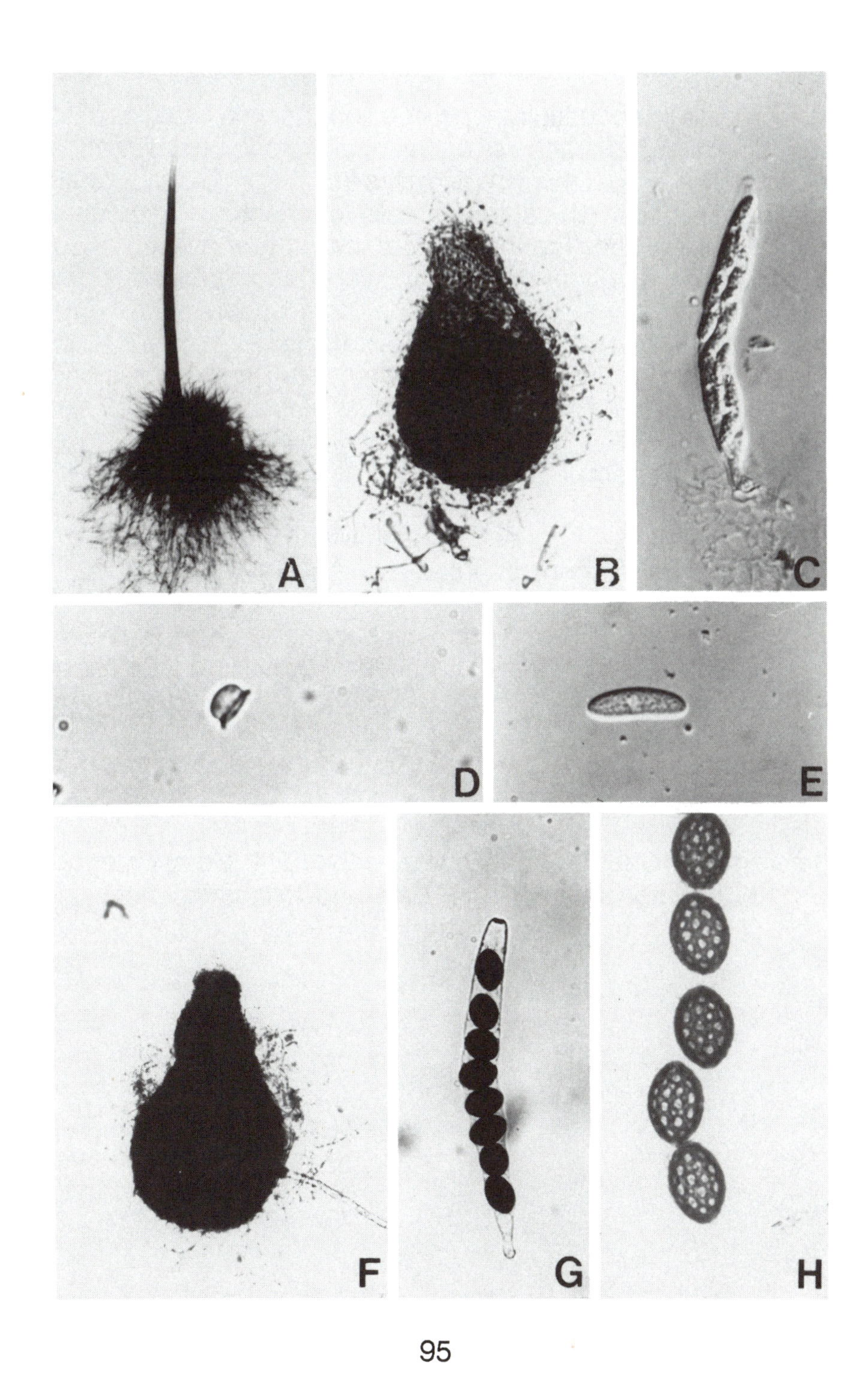
A
B
C
D
E
F
G
H

NEOCOSMOSPORA E. F. Sm.

Ascoma an ostiolate perithecium, single, globose to obpyriform, red to orange or occasionally brown; ostiolar neck short, lined with periphyses. Ascomal wall pseudoparenchymatous, outer cells pigmented, inner cells hyaline. Centrum containing apical paraphyses. Asci unitunicate, cylindrical or rarely clavate, short-stalked, apex undifferentiated or with an indistinct nonamyloid, ring, 8-spored. Ascospores 1-celled, uniseriate in ascus or occasionally biseriate, pale brown, with thick walls and roughened surface. (Fig. 5F-G).

Anamorph: *Acremonium.*

Habitat: In soil, often associated with roots, especially in tropical or subtropical areas.

Representative species: *Neocosmospora vasinfecta* E. F. Sm. (Anam. *Acremonium* sp.), cause of stem and root rot of legumes.

Comments: In *Pseudonectria* Seaver the ascospores are thin-walled, smooth, and elliptical.

References: Cannon and Hawksworth, 1984; Domsch et al., 1980; Matsushima, 1975; Ueda and Udagawa, 1983.

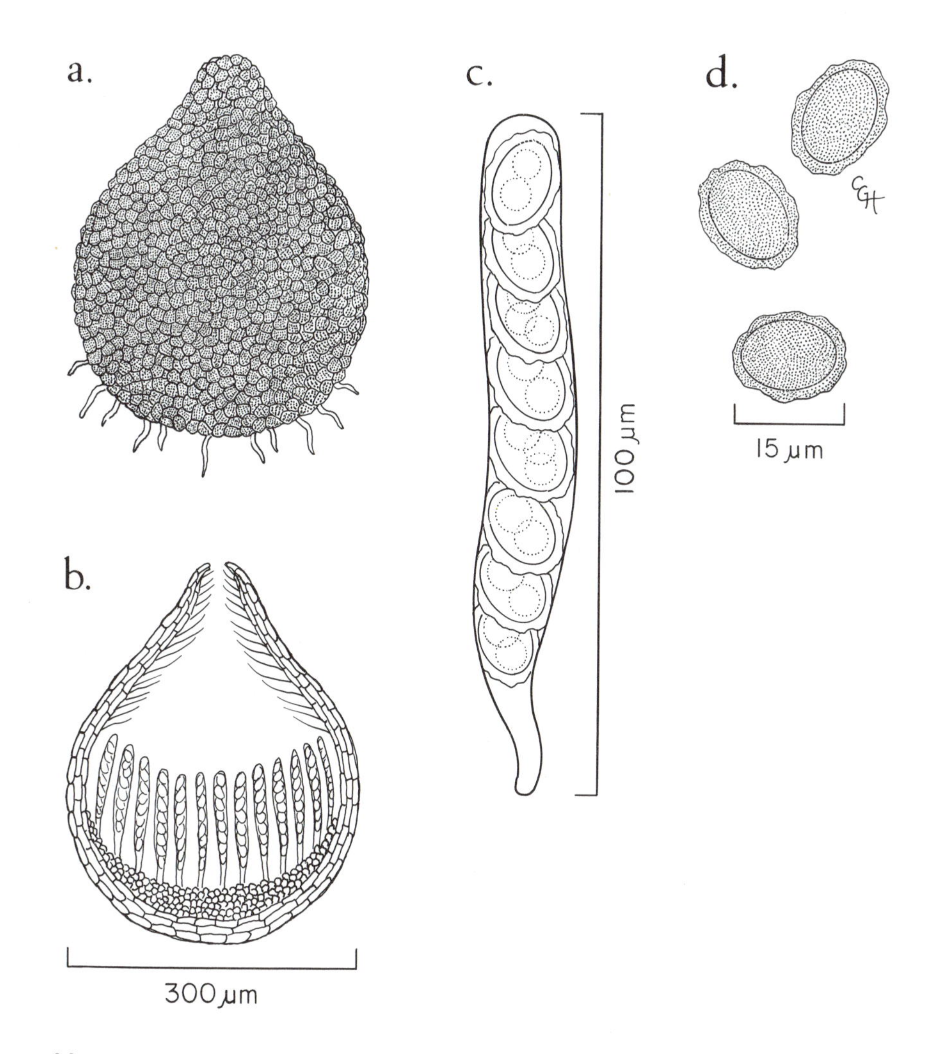

Neocosmospora tenuicristata. **a.** Exterior of perithecium. **b.** Section through perithecium with asci. **c.** Ascus with ascospores. **d.** Mature ascospores.

LOPHOTRICHUS R. K. Benjamin

Ascoma perithecioid, ostiolate, partially or entirely immersed in substrate, globose, with ostiolar neck short or long and cylindrical, clothed with dark brown hairs of various types. Ascomal wall transluscent when young, membranaceous and brittle when mature. Centrum pseudoparenchymatous, with scattered asci. Asci subglobose to broadly clavate, short-stalked, with evanescent walls, 8-spored. Ascospores 1-celled, limoniform or oval, light brown, with germ pore at each end.

Anamorph: None reported.

Habitat: Occurring on dung.

Representative species: *Lophotrichus ampullus* R. K. Benjamin

Comments: *Kernia* Nieuwland differs in having nonostiolate ascomata.

References: Ames, 1961; Benjamin, 1949; IM 11:C168.

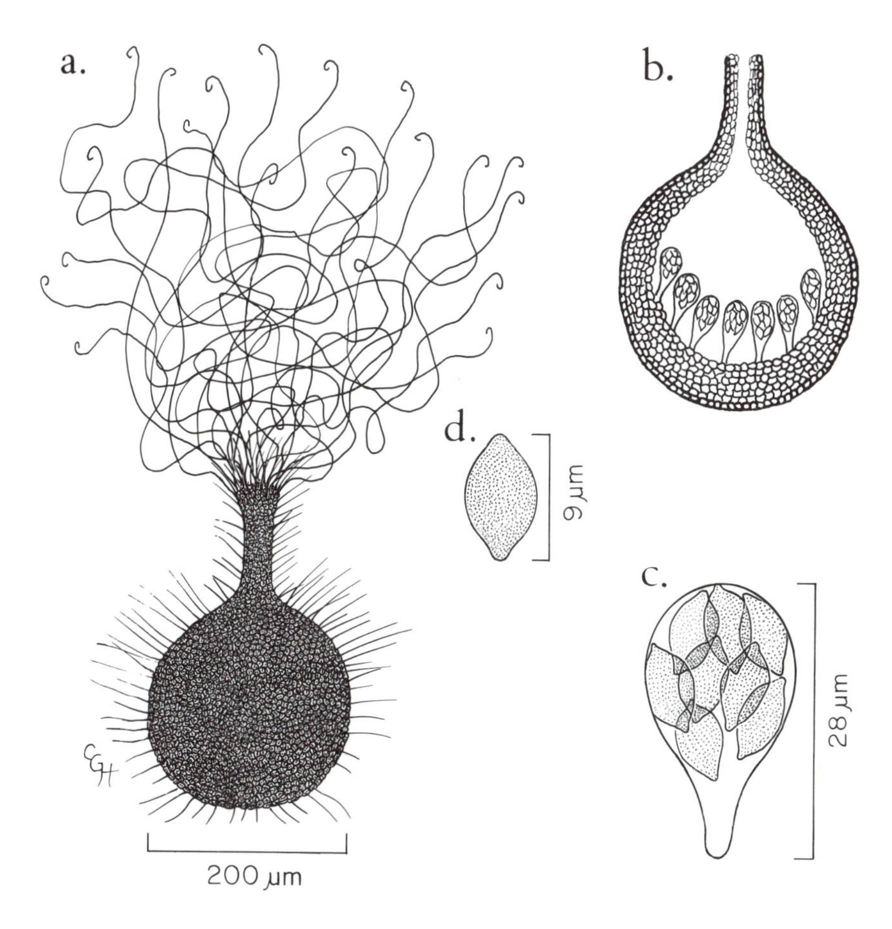

Lophotrichus ampullus. **a.** Exterior of perithecium. **b**. Section through perithecium with asci and ascospores. **c**. Ascus with ascospores. **d**. Mature ascospore.

CHAETOMIUM **Kunze:Fr.**

Ascoma an ostiolate perithecium, superficial, single to gregarious, globose, subglobose, elongated obpyriform, or vase-shaped, brown, clothed with hairs of various shapes; hairs brown, straight, flexuous, simple or dichotomously branched, coiled or arcuate, thick- or thin-walled, septate or aseptate, smooth or variously roughened. Ascomal wall translucent, membranaceous, pseudoparenchymatous or prosenchymatous. Paraphyses present or lacking. Asci clavate, linear to cylindrical, stalked, 4-8-spored, with evanescent walls. Ascospores 1-celled, light olive-brown to dark brown, with germ pore at one or both ends, often limoniform and apiculate, but also unbonate, almond-shaped, globose to subglobose, cymbiform or triangular, smooth, often pushed out of ostiole in a cirrhus.

Anamorphs: *Acremonium*-like, *Botryotrichum, Scytalidium*, and *Sporothrix*-like.

Habitat: Common on cellulosic substrates, also on seeds and in soil and dung.

Representative species: *Chaetomium globosum* Kunze : Fr., common on a wide range of substrates.

Comments: Species with straight, setose hairs are placed in *Farrowia* D. Hawksworth; in *Achaetomium* Rai, Tewari & Mukerji the ascospores are dark brown and opaque, the ascomal wall is thicker, and the hairs are short and inconspicuous.

References: Ames, 1961; Arx et al., 1984, 1986; Cannon, 1986; Domsch et al., 1980; Ellis and Ellis, 1985; Ellis and Everhart, 1892; Matsushima, 1971, 1975; Seth, 1972; Skolko and Groves, 1948, 1953; Udagawa, 1960; IM 1:C2.

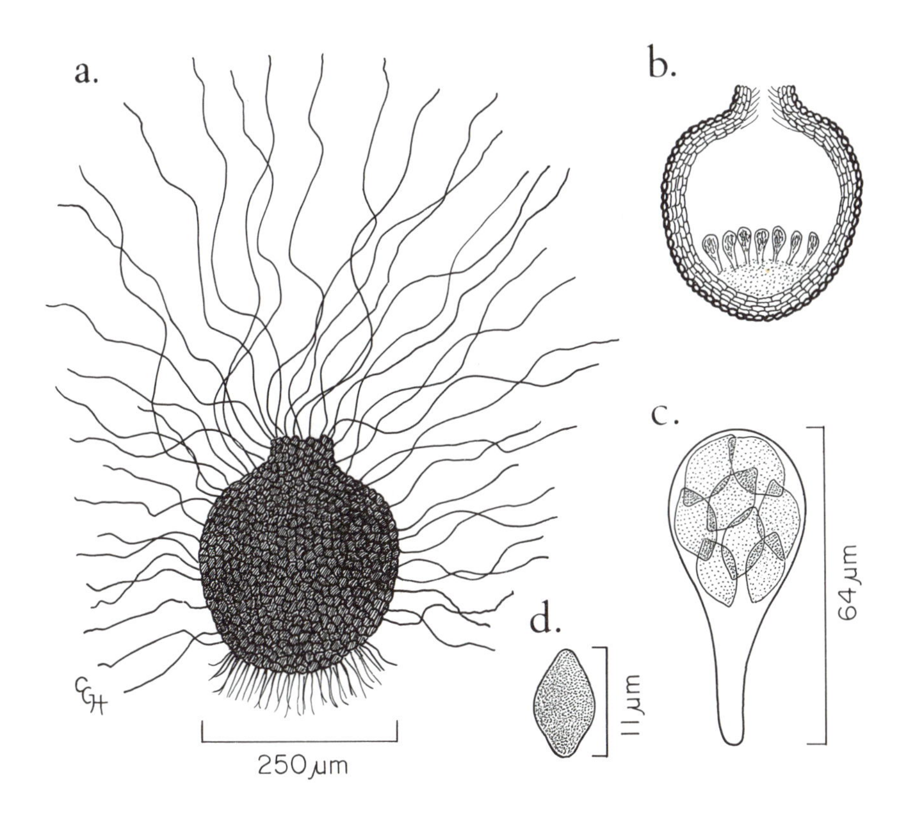

Chaetomium globosum **a.** Exterior of perithecium. **b**. Section through perithecium with asci. **c**. Ascus with ascospores. **d**. Mature ascospore.

ASCOTRICHA Berk.

Ascoma an ostiolate perithecium, dark-brown to black, formed singly on substrate, obpyriform to subglobose or globose with a short neck; ascomal wall pseudoparenchymatous, 2-3 cells thick, becoming carbonaceous at maturity. Ascoma clothed with brown to black hairs; terminal hairs arising from ostiolar neck, usually abundant, geniculate or curved, but never spirally coiled, simple or sympodially branched, with short, hyaline to light brown, sterile cells. Lateral hairs arising from sides of perithecium, like terminal hairs, but sometimes lacking. Young centrum with filiform lateral and hymenial paraphyses. Asci unitunicate, cylindrical, not blueing in iodine, not differentiated at apex, deliquescing at maturity, 8-spored. Ascospores pale brown to dark brown, arranged uniseriately in ascus, 1-celled, ellipsoidal, sometimes laterally compressed, with a germ slit.

Anamorph: *Dicyma.*

Habitat: Occurring on cellulosic materials, such as paper, wood, cloth, plant debris, and seeds.

Representative species: *Ascotricha chartarum* Berk. (Anam. *Dicyma ampullifera* Boul.), common on paper and plant materials.

Comments: *Coniochaeta* (Sacc.) Cooke differs in lacking perithecial hairs.

References: Ames, 1963; Hawksworth, 1971; Roberts et al., 1984; IM 13:C177.

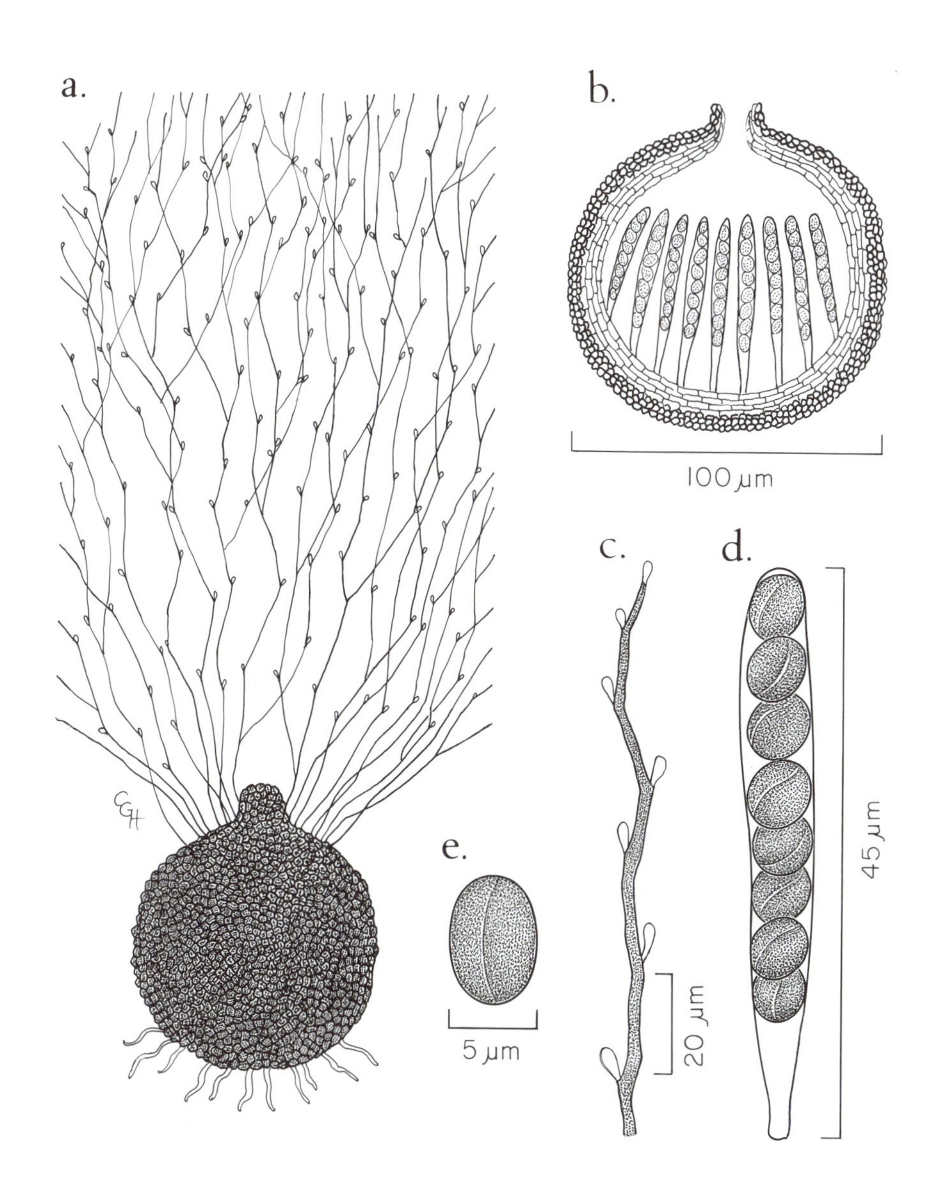

Ascotricha xylina. **a.** Exterior of perithecium. **b**. Section through perithecium with asci. **c**. Perithecial hair with sympodial branching and sterile cells. **d**. Ascus with ascospores. **e**. Ascospore with germ slit.

THIELAVIA **Zopf**

Ascoma a nonostiolate, globose perithecium, glabrous, setose or tomentose; ascomal wall brown, composed of flattened cells (textura epidermoidea). Asci in a fascicle or layer, ellipsoidal, saccate or cylindrical, rarely clavate, with a thin wall that is evanescent at maturity, 4-8-spored. Ascospores 1-celled, dark-brown, fusiform, ellipsoidal, obovate or clavate, with single germ pore at one end.

Anamorph: *Myceliophthora.*

Habitat: In soil, dung, plant debris and seeds.

Representative species: *Thielavia terricola* (Gilman & Abbott) Emmons, a common soil species.

Comments: The genus *Corynascus* Arx has been erected for species with a germ pore at each end of the spore and a *Chrysosporium* anamorph.

References: Arx, 1975a; Booth, 1961; Domsch et al., 1980; Matsushima, 1971, 1975; IM 9:C129

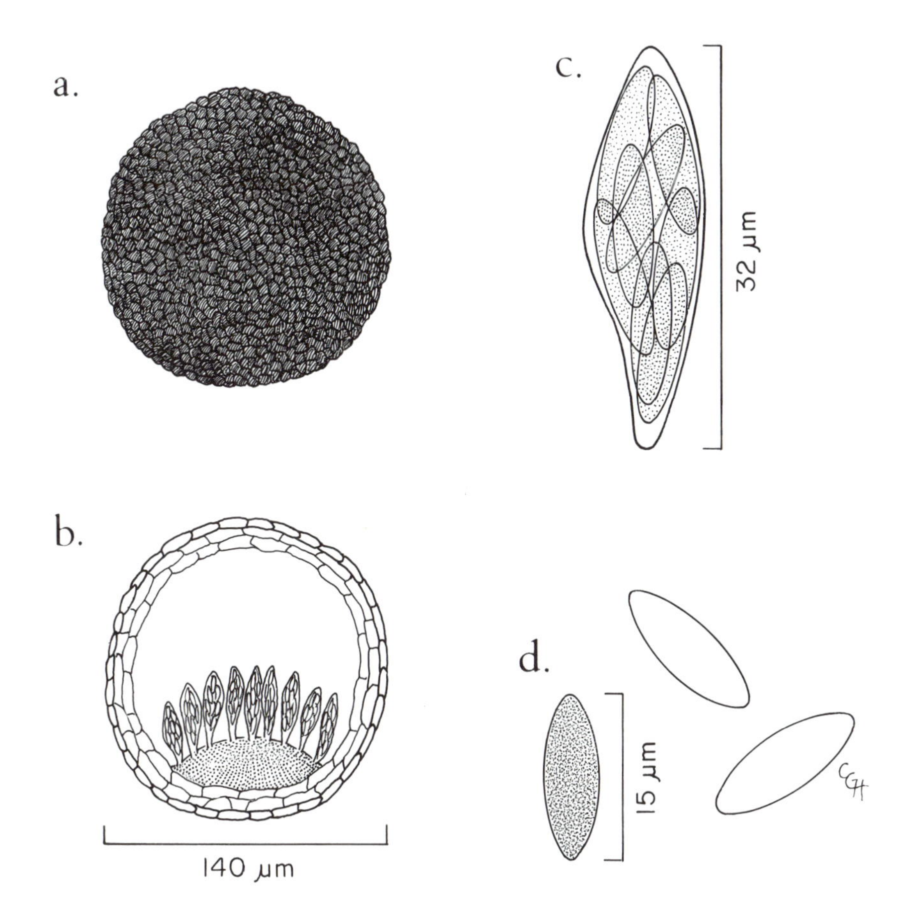

Thielavia terricola **a.** Exterior of perithecium. **b**. Section through perithecium with asci. **c**. Ascus with ascospores. **d**. Mature ascospores.

SORDARIA Ces. & De Not.

Ascoma an ostiolate, obpyriform, perithecium, nonstromatic, dark-brown to black, glabrous. Wall opaque, composed of 3-4 layers, cells of outer layer large, brown to black, rounded to angular, with slightly thickened walls, cells of middle layer large and hyaline, angular to flattened toward the interior, cells of innermost layer thin-walled, hyaline and flattened. Centrum with pseudoparenchyma lining wall and basal pseudoparenchyma bearing paraphyses; paraphyses in young ascomata filamentous, becoming vesiculose in older ascomata, forming before the asci and often disintegrating early. Asci unitunicate, cylindrical, with truncate apex, with distinct non-amyloid apical ring, short-stipitate, 8-spored, with spores arranged uniseriately. Ascospores dark brown, 1-celled, broadly fusiform to ovoid or subglobose, smooth, with basal germ pore and gelatinous sheath. (Fig. 7A-C).

Anamorph: None, but a phialidic stage is formed in some species.

Habitat: Mainly coprophilous, but also on seeds and in soil.

Representative species: *Sordaria fimicola* (Roberge ex Desmaz.) Ces. & De Not.

Comments: In *Neurospora* Shear & B. O. Dodge and *Gelasinospora* Dowding the ascospores lack a gelatinous sheath; in *Neurospora* the ascospores are longitudinally striate and in *Gelasinospora* they are pitted or reticulate.

References: Cain and Groves, 1948; Domsch et al., 1980; Ellis and Everhart, 1892; Lundqvist, 1972; Matsushima, 1971, 1975; Moreau, 1953; IM 1:C1.

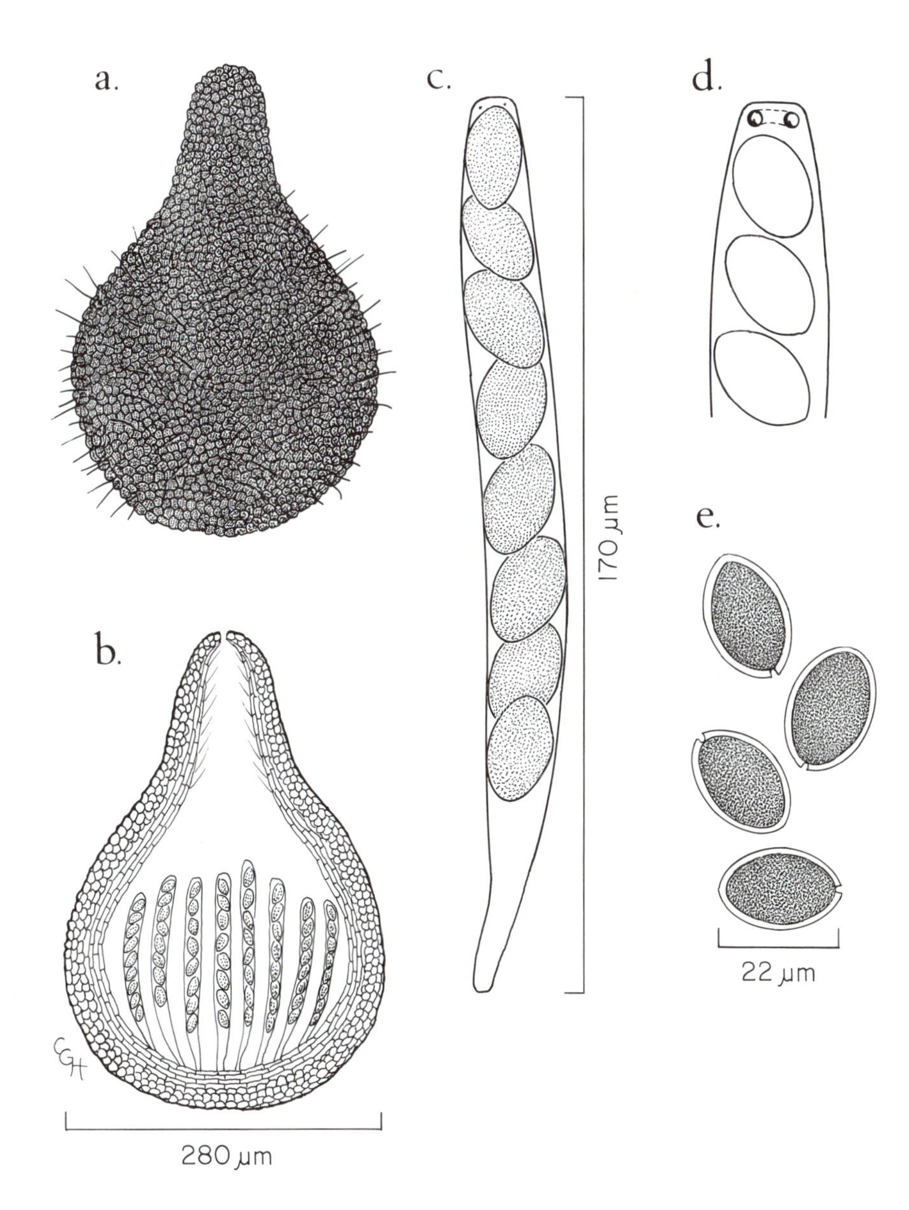

Sordaria fimicola. **a.** Exterior of perithecium. **b**. Section through perithecium with asci. **c**. Ascus with ascospores. **d**. Ascus with apical ring. **e**. Mature ascospores with gelatinous sheath.

NEUROSPORA Shear & B. O. Dodge

Ascoma an ostiolate perithecium, single, superficial, or occasionally gregarious and partially immersed in substrate; perithecium globose or obpyriform, brown, glabrous or with scattered hyphae; ostiolar neck papillate to elongated, lined with periphyses. Perithecial wall several layers thick, outer cells angular, brown and translucent, inner cells flattened and hyaline to subhyaline. Centrum containing paraphyses when young. Asci unitunicate, cylindrical truncate at apex with a highly refractive, nonamyloid ring, stipitate, 3-8-spored. Ascospores 1-celled, ellipsoidal, dark brown, with longitudinal striations on surface and an indistinct germ pore at both ends.

Anamorph: *Chrysonilia (Monilia*).

Habitat: On organic materials, especially those with high sugar content.

Representative species: *Neurospora sitophila* Shear & B. O. Dodge [Anam. *Chrysonilia sitophila* (Mont.) Arx], cause of red mold of bakery products.

Comments: In *Gelasinospora* Dowding the ascospores are pitted or reticulate; in *Sordaria* Ces. & De Not. they are smooth and surrounded by a gelatinous sheath.

References: Arx and Müller, 1954; Dennis, 1981; Matsushima, 1971, 1975; Moreau, 1953; IM 9:C135.

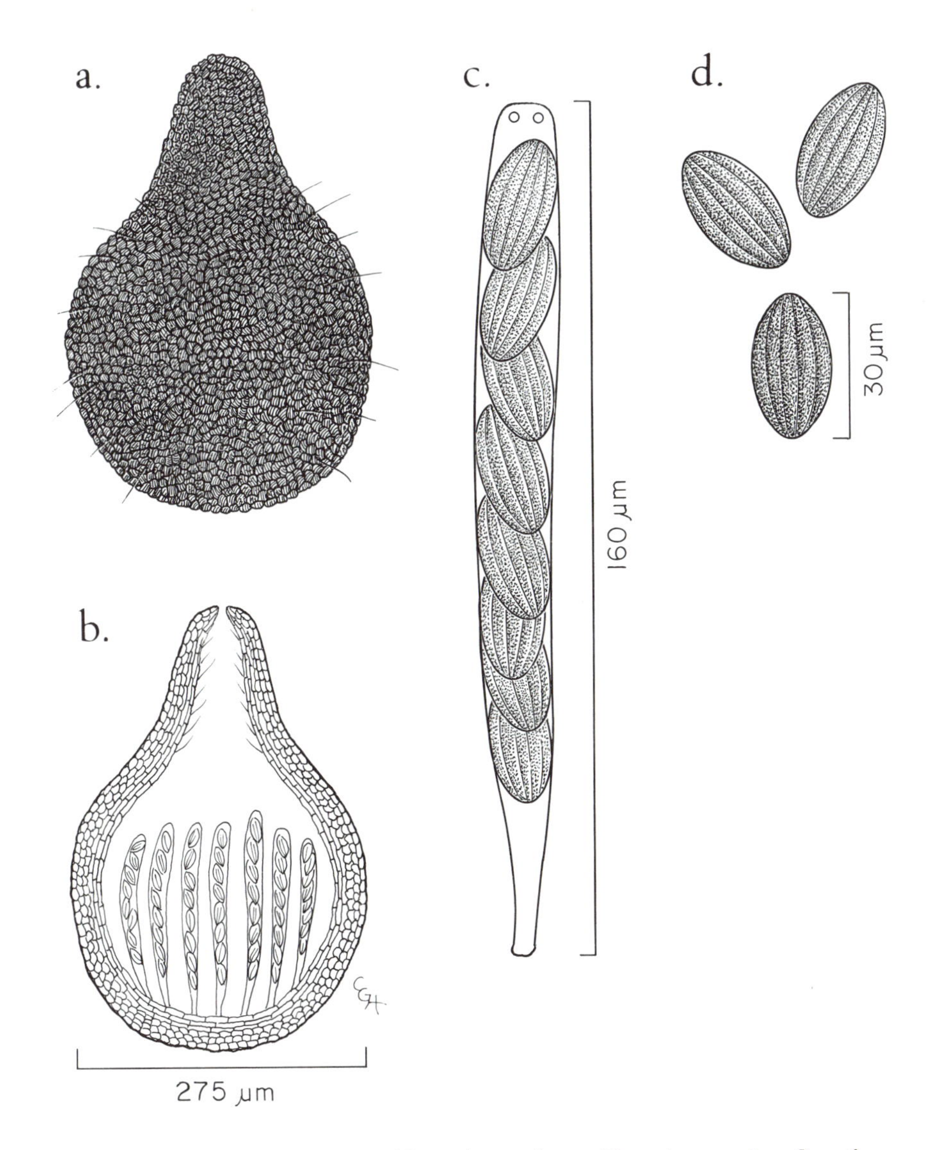

Neurospora crassa. **a.** Exterior of perithecium. **b**. Section through perithecium with asci. **c**. Ascus with ascospores. **d**. Longitudinally striate ascospores.

GELASINOSPORA Dowding

Ascoma a nonstromatic, ostiolate, obpyriform perithecium, often covered with dark brown, short mycelial hairs; wall 3-layered, opaque; cells of outer layer pseudoparenchymatous, dark brown, with slightly thickened walls, cells of middle layer large, hyaline, angular, merging to flattened toward the interior, cells of inner layer hyaline, thin-walled, flattened. Centrum in young ascomata containing pseudoparenchyma lining interior of wall and a basal mass of pseudoparenchyma cells bearing paraphyses; paraphyses in young ascomata filamentous, becoming vesiculose when old. Asci unitunicate, cylindrical, with truncate apex and distinct, nonamyloid apical ring, short-stipitate, 4-8-spored, spores uniseriate in ascus. Ascospores 1-celled, olivaceous when young, then brown to blackish, ellipsoid to oval or subglobose, with a pitted or reticulate surface, with 1-2 germ pores, often somewhat inaequilateral. (Fig. 3F-H).

Anamorph: None reported.

Habitat: On seeds, in soil and dung.

Representative species: *Gelasinospora calospora* V. (Mouton) C. Moreau & Mme. Moreau.

Comments: The characteristic pits are most easily observed in young ascospores before they become brown; fully pigmented ascospores may be mistaken for *Sordaria* Ces. & De Not., which has smooth ascospores that are surrounded by a gelatinous sheath. In *Neurospora* Shear & B. O. Dodge the ascospores are longitudinally striate.

References: Arx, 1982; Cailleux, 1971; Cain, 1950; Domsch et al., 1980; Ellis and Ellis, 1985; Lundqvist, 1972; Matsushima, 1971, 1975; IM 19:C118.

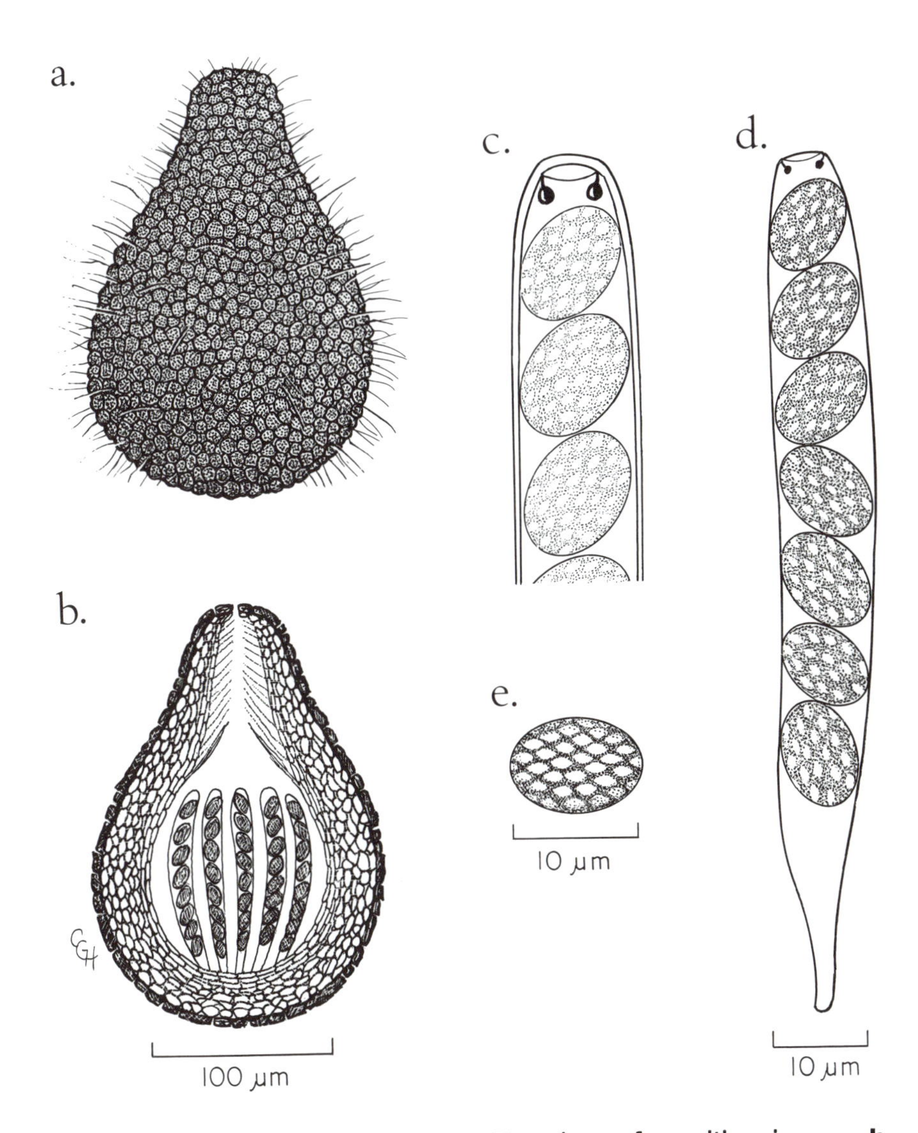

Gelasinospora retispora. **a.** Exterior of perithecium. **b**. Section through mature perithecium with asci and ascospores. **c**. Ascus apex. **d**. Ascus with ascospores. **e**. Mature ascospore.

CONIOCHAETA (Sacc.) Cooke

Ascoma an ostiolate perithecium, single or aggregated, superficial or partially immersed in substrate, subglobose to obpyriform, dark brown, glabrous or bearing brown setae or stiff hairs. Ostiole papillate, lined with periphyses. Perithecial wall 2-layered, outer layer composed of brown, thick-walled, angular cells, the inner layer of thin-walled, hyaline, flattened cells. Centrum containing paraphyses. Asci unitunicate, cylindrical, stalked, 8-spored, uniseriate. Ascospores 1-celled, brown, ellipsoidal, with a longitudinal germ slit.

Anamorph: *Phialophora.*

Habitat: On wood and herbaceous stems.

Representative species: *Coniochaeta ligniaria* (Grev.) Cooke.

Comments: *Coniochaetidium* Malloch & Cain differs in having nonostiolate ascomata. In *Ascotricha* Berk. the ascomata bear sympodially branched hairs with sterile cells.

References: Arx and Müller, 1954; Dennis, 1981; Ellis and Ellis, 1985; Furuya and Udagawa, 1977; Matsushima, 1971, 1975; IM 2:C25

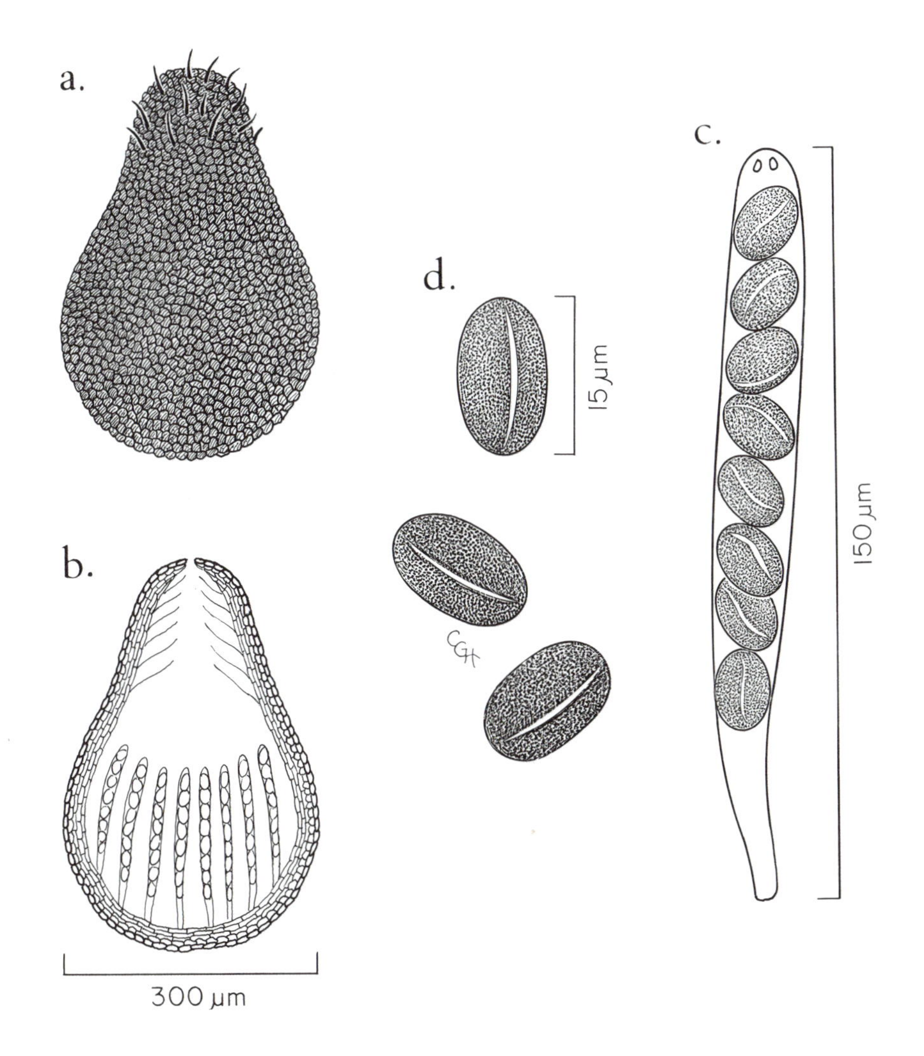

Coniochaeta ligniaria. **a.** Exterior of perithecium. **b**. Section through perithecium with asci. **c**. Ascus with ascospores. **d**. Mature ascospores with germ slit.

ROSELLINIA De Not.

Ascoma an ostiolate perithecium, borne on a feltlike subiculum of dark hyphae, single or gregarious, often covering the surface of subiculum; perithecia globose or broadly obpyriform, mostly over 500 µm diameter, black, glabrous, often seated on a definite hypostroma composed of angular, thick-walled cells. Ostiolar neck papillate or conical, ostiole lined with periphyses. Perithecial wall composed of several layers of flattened cells, outer cells thick-walled and dark brown, inner cells subhyaline or hyaline. Asci unitunicate, cylindrical, long-stalked, with an amyloid apical apparatus, lining inside of perithecial wall, 8-spored. Ascospores 1-celled, dark brown, ellipsoidal, often laterally compressed, with a longitudinal germ slit.

Anamorphs: *Dematophora, Nodulisporium*, and *Sporothrix.*

Habitat: On wood, roots or similar substrates.

Representative species: *Rosellinia necatrix* Prill., cause of root rot of numerous tree species.

Comments: *Hypoxylon* Bull.: Fr. and *Xylaria* J. Hill ex Schrank differ in having the perithecia immersed in a fungal stroma.

References: Arx and Müller, 1954; Dargan and Thind, 1979; Ellis and Ellis, 1985; Ellis and Everhart, 1892; Francis, 1985; Martin, 1967; Petrini and Petrini, 1985; Saccas, 1956; BK 352; CMI 351-354; IM 4:C42.

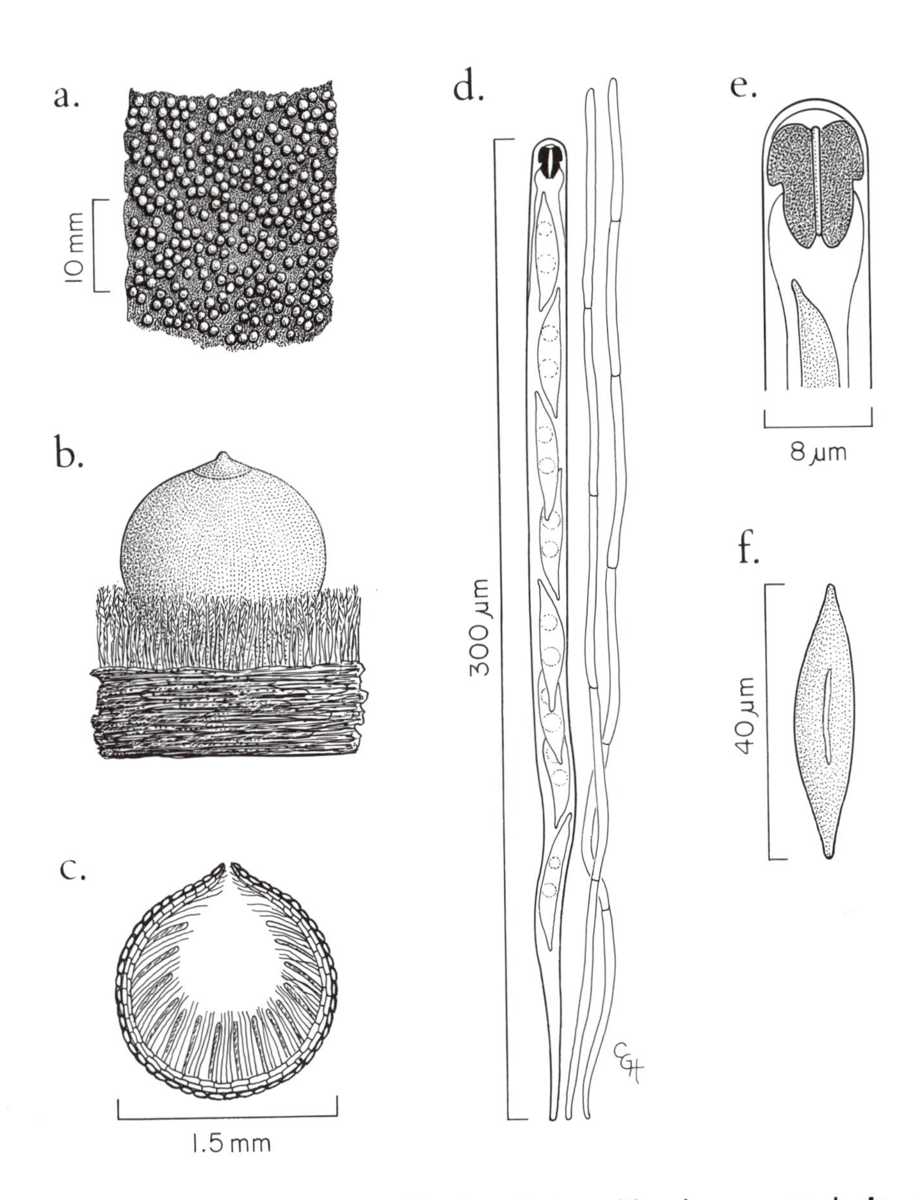

Rosellinia necatrix **a.** Habit showing perithecia on wood. **b.** Exterior of perithecium seated on subiculum. **c**. Section through perithecium with asci. **d**. Paraphyses and ascus with ascospores. **e**. Apex of ascus. **f**. Mature ascospore.

PODOSORDARIA Ellis & Holw.

Ascoma an ostiolate perithecium, formed in a cluster at the tip of an erect, brown or black stroma with a cylindrical stalk and a subglobose head; perithecia subglobose, black or brown, erumpent or immersed, with papillate ostioles. Perithecial wall brown, membranaceous, pseudoparenchymatous, usually 2-layered. Paraphyses abundant, filiform to tubular, septate. Asci unitunicate, cylindrical to subpyriform, short-stalked, with a distinct apical ring that stains blue in iodine, 8-spored. Ascospores 1-celled, obliquely uniseriate to biseriate in ascus, ellipsoidal, dark brown, smooth, with a germ slit and a thin gelatinous sheath.

Anamorph: Unnamed; conidia form on stroma.

Habitat: Growing in dung.

Representative species: *Podosordaria leporina* (Ellis & Everh.) Dennis.

Comments: *Poronia* Willd. differs in having a disc-shaped, stalked stroma with completely immersed perithecia.

References: Dennis, 1981; Furuya and Udagawa, 1977; Krug and Cain, 1974; Martin, 1970.

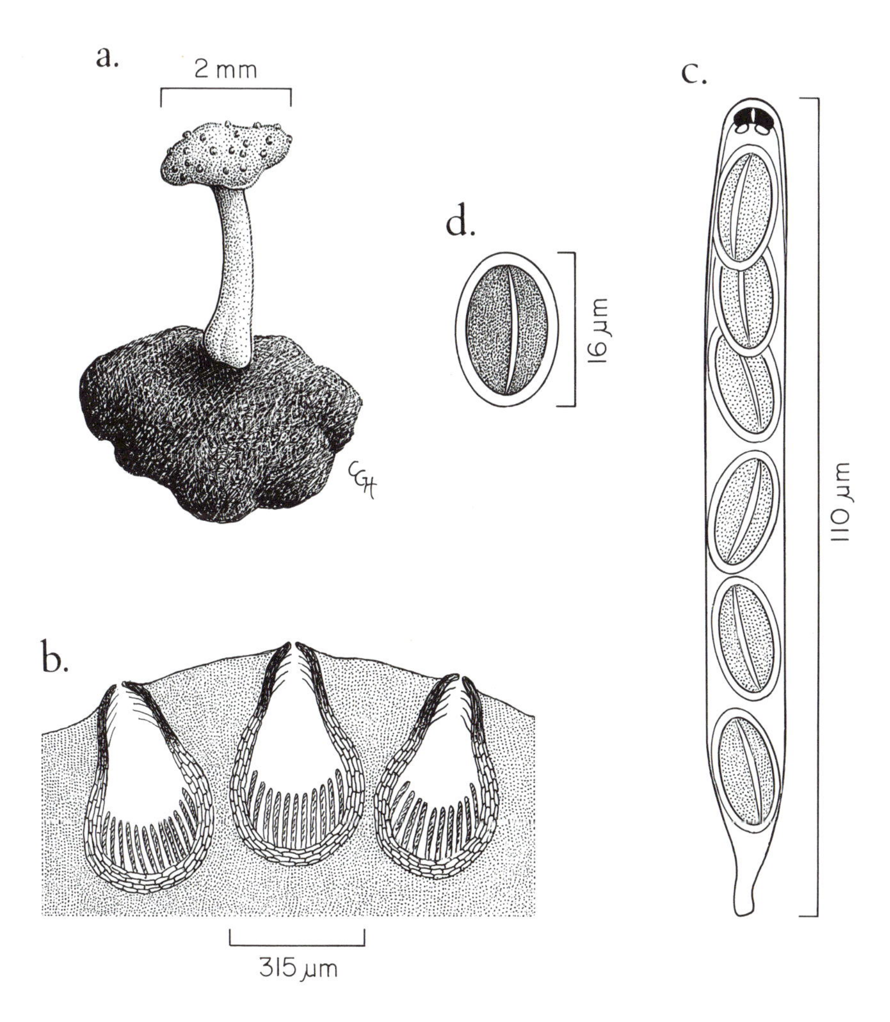

Podosordaria leporina **a.** Perithecial stroma on dung. **b.** Section through portion of head of perithecial stroma with perithecia and asci. **c.** Ascus with ascospores. **d.** Mature ascospore.

HYPOXYLON **Bull.:Fr.**

Ascoma an ostiolate perithecium immersed in a stroma composed entirely of fungal tissue; stroma convex, applanate, pulvinate, or subglobose, determinate or widely effused, erumpent from substrate, fleshy to leathery or woody and carbonaceous, interior dark, not zonate, surface black, brown, reddish-brown, or gray. Perithecia forming a single layer in periphery of stroma, obpyriform, base subglobose or angular; ascomal wall composed of dark, thick-walled cells; cells pseudoparenchymatous or flattened, interior cells hyaline and thin-walled. Centrum containing filamentous paraphyses. Asci unitunicate, cylindrical, with a distinct amyloid apical apparatus, forming a layer along base and sides of perithecium, 8-spored. Ascospores 1-celled, light to dark brown, subglobose to ellipsoid, smooth, with germ slit.

Anamorphs: *Nodulisporium* and *Geniculosporium.*

Habitat: Occuring on bark or wood, occasionally parasitic.

Representative species: *Hypoxylon atropunctatum* (Schwein.: Fr.) Cooke on oak *(Quercus* spp.).

Comments: In *Daldinia* Ces. & De Not. the interior of the stroma is zonate; in *Penzigia* Sacc. it is white and not zonate.

References: Ellis and Ellis, 1985; Ellis and Everhart, 1892; Martin, 1967, 1968a,b,c; Miller, 1961; Petrini and Rogers, 1986; Whalley and Greenhalgh, 1973; Viégas, 1944; BK 338-344; CMI 356-359; IM 27:C468.

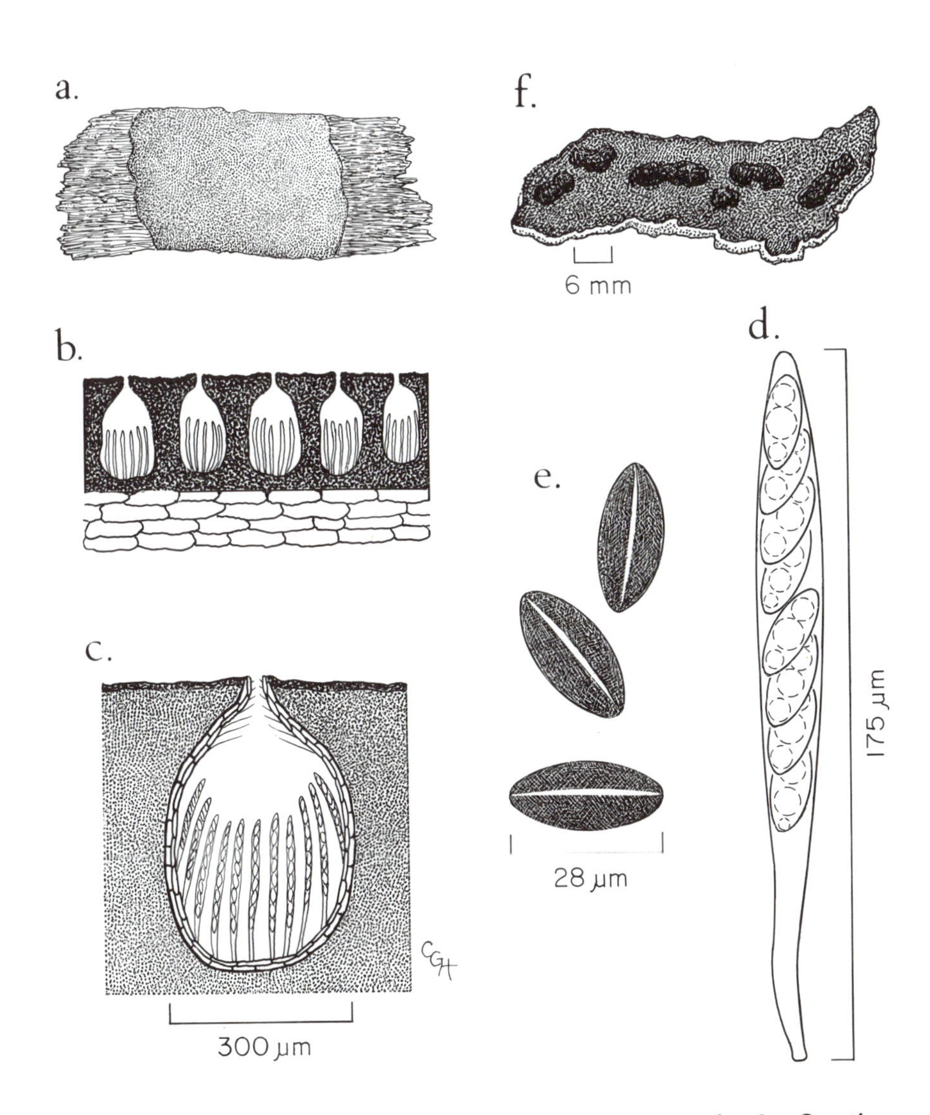

Hypoxylon atropunctatum. **a.** Stroma on wood. **b.** Section through stroma with perithecia. **c.** Section through perithecium with asci. **d.** Ascus with ascospores. **e.** Mature ascospores with germ slit. **f.** *Hypoxylon rubiginosum.* Pulvinate stromata.

DALDINIA Ces. & De Not.

Ascoma an ostiolate perithecium that is immersed in a stroma; stroma black, carbonaceous, globose or subglobose to hemispherical, up to 3 cm in diameter, composed entirely of fungal tissue, seated directly on substrate or with a short stalk; interior zonate, consisting of concentric light and dark layers. Perithecia globose to ellipsoidal or obpyriform, immersed in periphery of stroma in a single layer; ostiolar necks lined with periphyses. Perithecial wall composed of flattened cells, exterior cells thick-walled and dark, inner cells thin-walled and hyaline. Centrum containing filamentous paraphyses. Asci unitunicate, cylindrical, with a short stalk and amyloid apical apparatus. Ascospores 1-celled, dark brown, uniseriate, elliptical-fusiform, inaequalateral, with germ slit.

Anamorph: *Nodulisporium.*

Habitat: On dead wood.

Representative species: *Daldinia concentrica* (Bolton:Fr.) Ces. & De Not.

Comments: In *Hypoxylon* Bull.: Fr. the interior of the stroma is not zonate, and in *Rosellinia* De Not. the perithecia are associated with a basal subiculum.

References: Arx and Müller, 1954; Dennis, 1981; Ellis and Ellis, 1985; Ellis and Everhart, 1892; Martin, 1969; Pérez-Silva, 1973; Thind and Dargan, 1978; Viégas, 1944; BK 346; IM 26:C460.

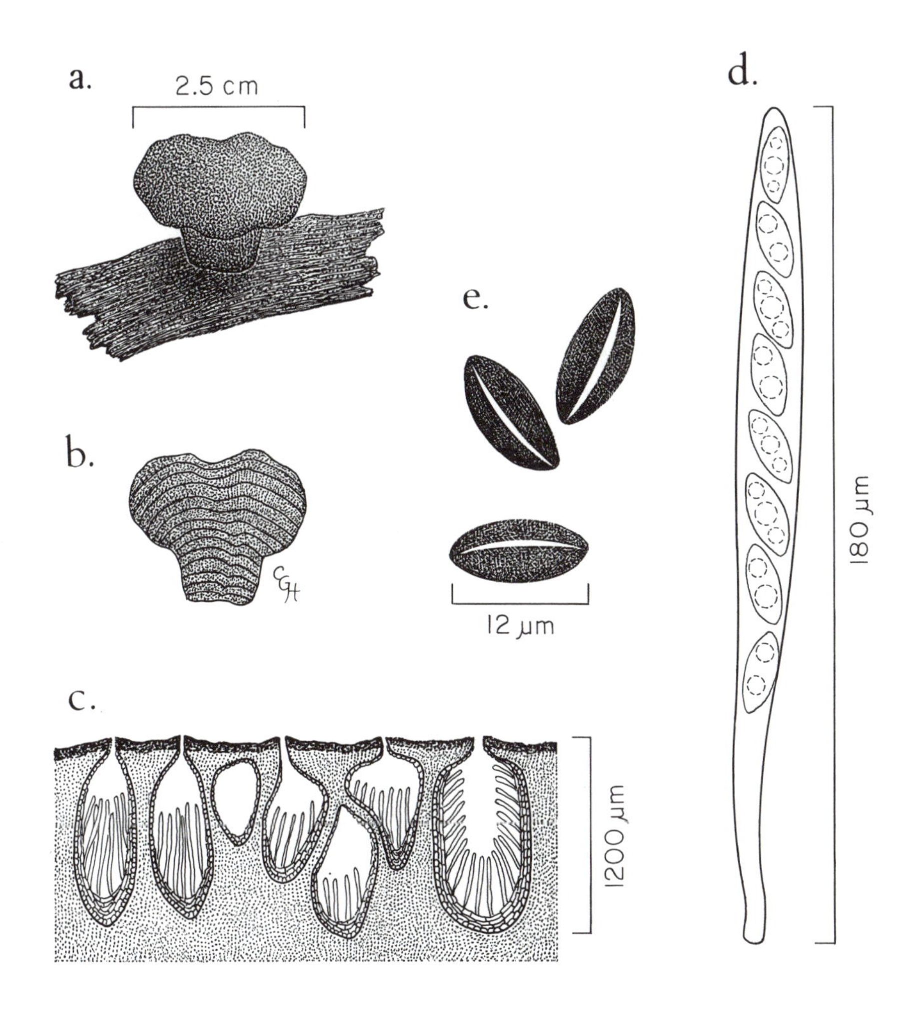

Daldinia concentrica. **a.** Perithecial stroma on wood. **b.** Section through stroma showing zonation. **c.** Section through periphery of stroma with perithecia. **d.** Ascus with ascospores. **e.** Mature ascospores with germ slit.

XYLARIA J. Hill ex Schranck

Ascoma an ostiolate perithecium formed in a stroma composed entirely of fungal tissue; stroma erect, up to 10 cm high, cylindrical, clavate or slender, simple or branched, terete to ellipsoidal in cross-section, corky or fleshy; exterior dark, composed of thick-walled cells, often roughened or covered with matted, dark brown, hairs; dark brown to dark gray or black, tip often whitish and conidiogenous; interior white or sometimes brownish, composed of hyaline, interwoven hyphae, center hollow in some species. Perithecia immersed, forming a layer on periphery of stroma, with slightly protruding ostiolar papillae, obpyriform to ellipsoidal, often laterally compressed by mutual pressure, ostiole lined with periphyses. Perithecial wall cells dark and flattened on exterior, subhyaline on interior. Paraphyses filamentous. Asci unitunicate, forming a layer along sides and base of perithecium, cylindrical to clavate, with distinct, amyloid apical apparatus, 4-8-spored. Ascospores 1-celled, dark brown, oval to ellipsoid, sometimes inaequilateral, with a longitudinal germ slit or a lateral germ pore.

Anamorph: *Nodulisporium.*

Habitat: On dead wood, but also often isolated from leaves and other cellulosic plant materials; a few species are pathogenic.

Representative species: *Xylaria polymorpha* (Pers.:Fr.) Grev., common on dead wood.

Comments: In *Kretzschmaria* Fr. the stroma is flat, crust-like and brittle with knobby fertile regions on the surface; in *Camillea* Fr. the stroma is erect and cylindrical and bears elongate perithecia beneath a fold at the apex of the stroma.

References: Dennis, 1956, 1958, 1981; Ellis and Ellis, 1985; Ellis and Everhart, 1892; Martin, 1970; Pérez-Silva, 1975; Poroca, 1986; Rogers, 1986; Rogers and Callan, 1986; Viégas, 1944; BK 347-351; CMI 355.

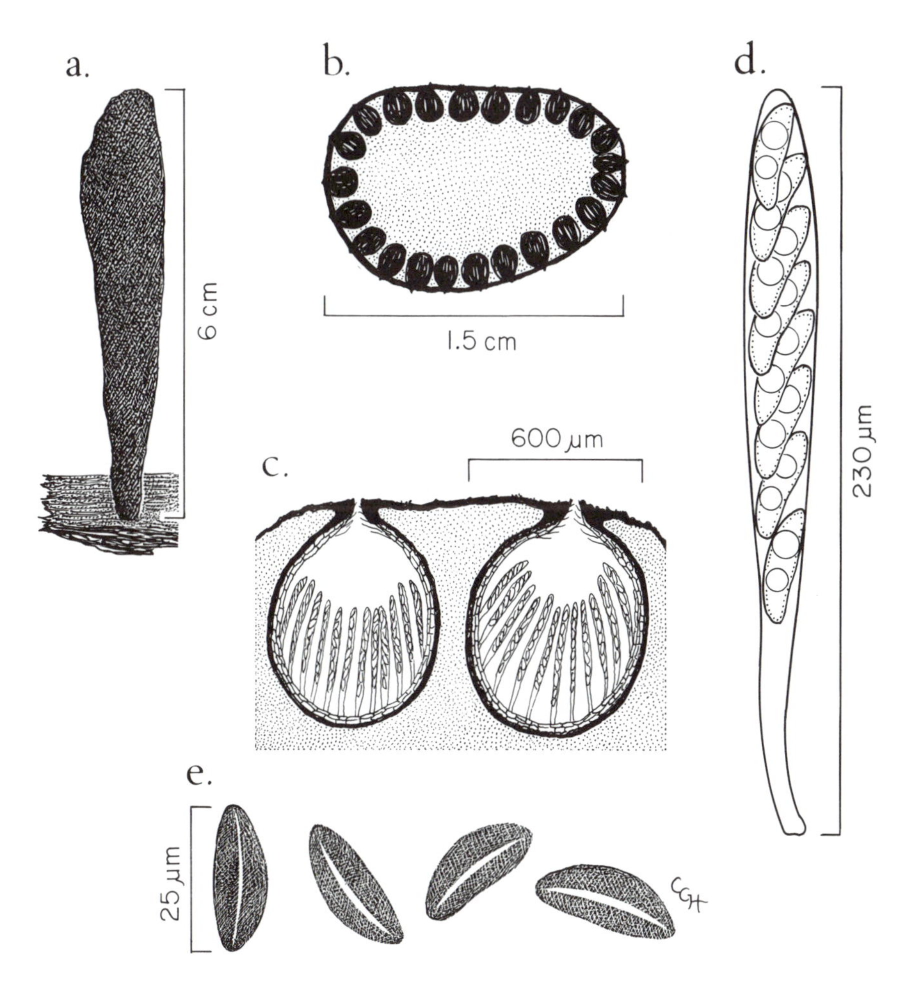

Xylaria polymorpha. **a.** Stroma on wood. **b.** Cross-section through stroma with perithecia. **c.** Close-up of perithecia with asci. **d.** Ascus with ascospores. **e.** Mature ascospores with germ slit.

ASCOBOLUS Pers.

Ascoma an apothecium, superficial or immersed, sessile, but sometimes with a short stalk, up to 30 mm diameter; receptacle subglobose, pyriform, obconical, cupulate or saucer-shaped, smooth, furfuraceus, villose or downy. Paraphyses slender, cylindrical, often embedded in mucus. Asci unitunicate, operculate, saccate-clavate or cylindric-clavate, with rounded or dome-shaped apex, protruding above surface of hymenium when mature, amyloid in some species, 4 or 8-spored. Ascospores 1-celled, subglobose to elliptical or oval, thick-walled, sometimes with a gelatinous sheath, ornamented with pigment deposited externally on spore; spores smooth or variously roughened, 2-3-seriate in ascus, ejected singly, purple, then brown in color.

Anamorph: None reported.

Habitat: Mostly coprophilous, but also on leaves, wood and soil.

Representative species: *Ascobolus furfuraceus* Pers.

Comments: In *Saccobolus* Boud. the ascospores are united in a cluster and are ejected together.

References: Brummelen, 1967; Dennis, 1981; Ellis and Ellis, 1985; Seaver, 1942; BK 111-114; IM 7:C104.

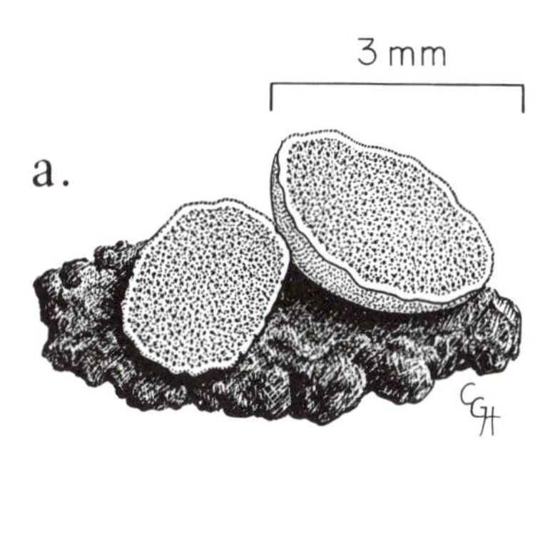

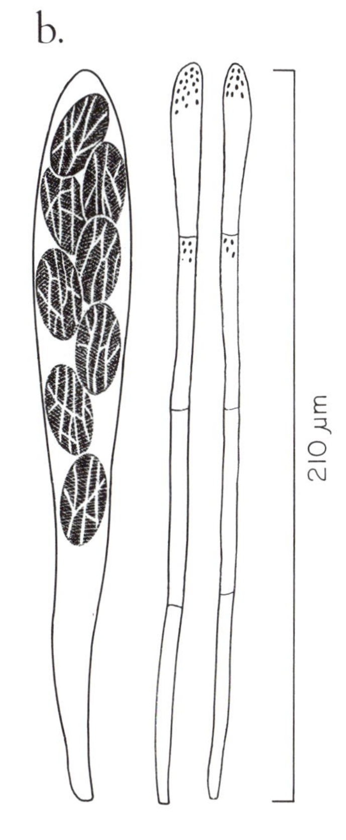

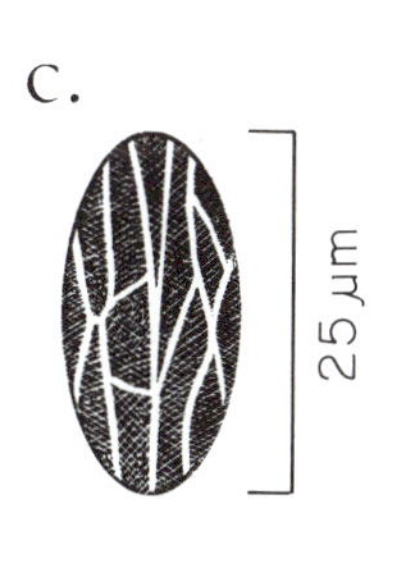

Ascobolus furfuraceus. **a.** Apothecia on dung. **b**. Paraphyses and ascus with ascospores. **c.** Mature ascospores.

TUBER P. Mich. ex A. Wigg.

Ascoma hypogeous, enclosed, globose to lobed or irregular in shape, fleshy or cartilaginous, exterior smooth or roughened, usually light to dark brown, but sometimes whitish, red or black. Interior of ascoma composed of a peripheral cortex (medullary excipulum) with an exterior layer (ectal excipulum) and a central gleba. Cortex often pseudoparenchymatous, sometimes prosenchymatous. Gleba consisting of convoluted, parallel, light and dark veins, the sterile venae externae and the fertile venae internae; tissue of veins interwoven, parallel or sometimes pseudoparenchymatous. Asci unitunicate, irregularly arranged between the veins, pyriform, ellipsoid or subglobose, short- or long-stalked, containing 1-8 ascospores. Ascospores 1-celled, yellow-brown to brown, variable in size, ellipsoid to globose, surface alveolate or spiny. (Fig. 5A-C).

Anamorph: None reported.

Habitat: In soil, often under trees; some species are known to form mycorrhizae.

Representative species: *Tuber melanosporum* Vittadini, the Perigord truffle.

Comments: In *Pachyphloeus* Tul. the asci are clavate.

References: Gilkey, 1939, 1954; BK 124-125.

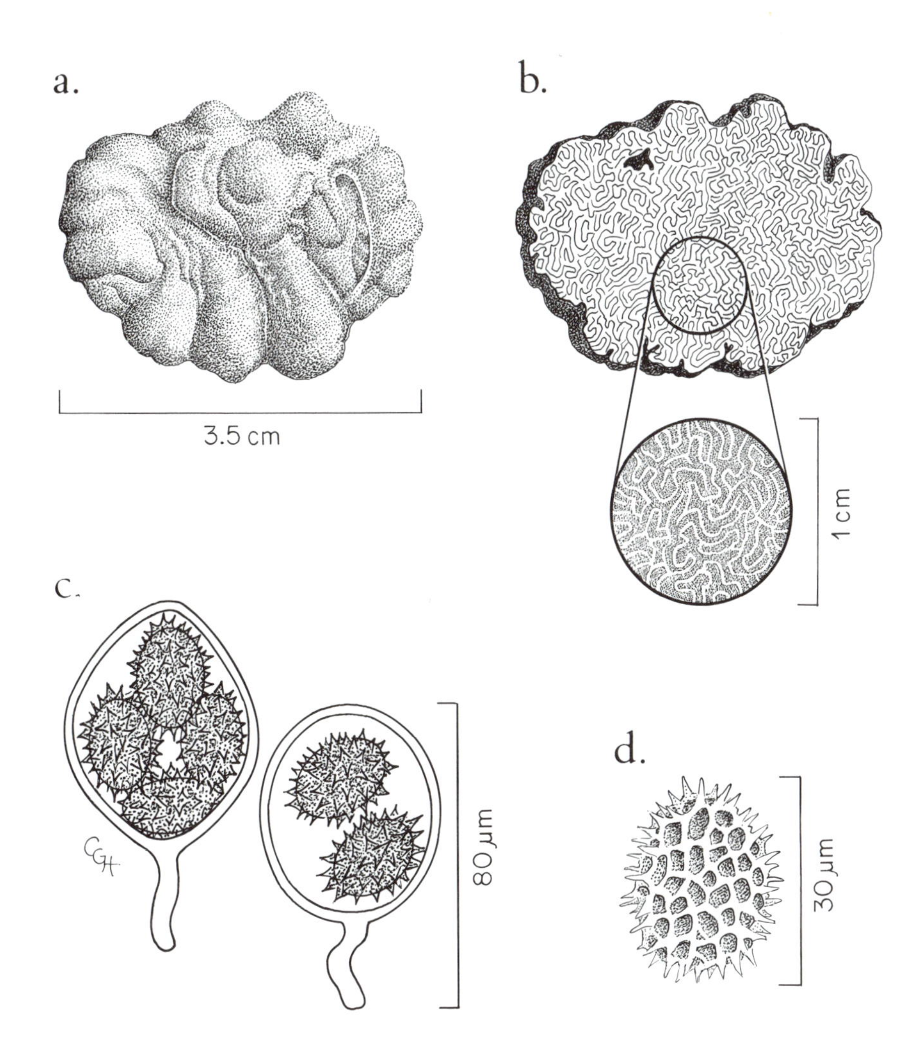

Tuber texense. **a.** Exterior of ascoma. **b.** Section through ascoma showing white sterile veins and brown fertile veins. **c.** Asci with ascospores. **d.** Mature ascospore with spines and reticulate surface pattern.

GNOMONIA Ces. & De Not.

Ascocarp an ostiolate perithecium, non- stromatic, brown, immersed in host tissue, globose to subglobose, usually with long ostiolar neck; ascomal wall composed of several layers of dark, compressed cells. Asci numerous, unitunicate, ellipsoid, with refractive, nonamyloid, apical ring, with deliquescent bases, lying free in perithecium at maturity, 2-many-spored. Ascospores hyaline, occasionally brownish in age, 2-celled, ellipsoid, fusoid, or cylindrical, septum median, or sub- or supra-median, sometimes developing additional septa in age, biguttulate, smooth, sometimes with appendages.

Anamorphs: *Cylindrosporella, Discula, Gloeosporium, Marssoniella,* and *Zythia.*

Habitat: Occurring on leaves and petioles of deciduous trees and on leaves and stems of herbaceous plants, usually saprobic, but some species parasitic.

Representative species: *Gnomonia comari* P. Karst. (Anam. *Zythia fragariae* Laibach), cause of leaf spot of strawberry.

Comments: *Gnomoniella* Sacc. is similar to *Gnomonia,* but it has 1-celled ascospores and usually a shorter ostiolar neck. In *Apiognomonia* Höhn. the ascospores are 2-celled, but they are apiosporous.

References: Barr, 1978; Dennis, 1981; Ellis and Ellis, 1985; Ellis and Everhart, 1892; Kobayashi, 1970; Monod, 1983; BK 371; CMI 737; IM 4:C36.

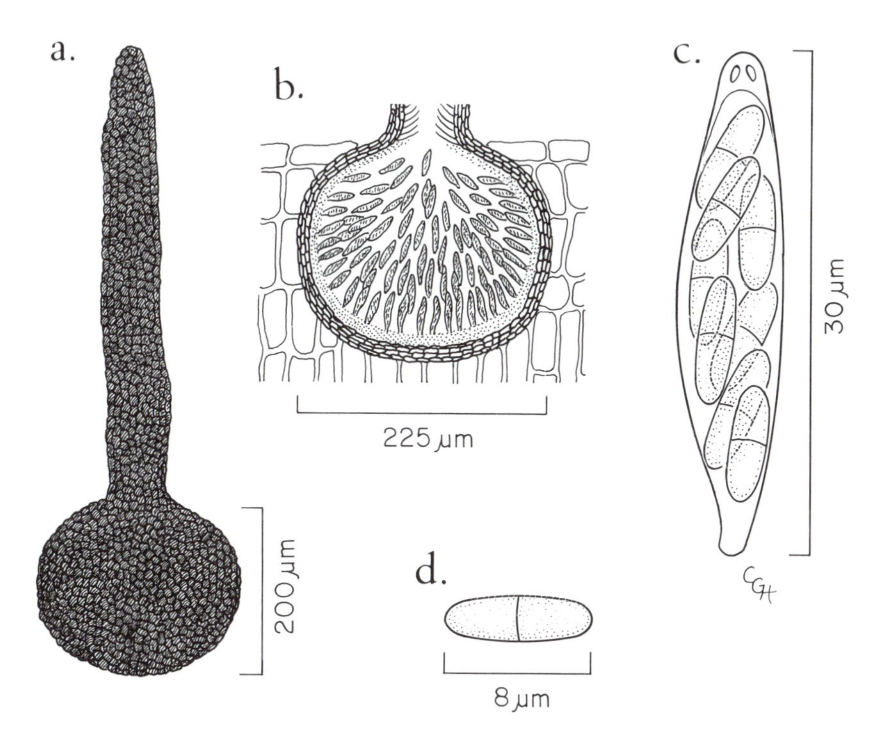

Gnomonia comari. **a.** Exterior of perithecium. **b.** Section through base of perithecium with asci. **c.** Ascus with ascospores. **d.** Mature ascospore.

STEGOPHORA Syd. & P. Syd.

Ascoma an ostiolate perithecium formed in host tissues beneath a black, crustose clypeus in the upper epidermis; perithecia upright, immersed, often several clustered beneath a common clypeus, ostioles short, erumpent through lower epidermis. Ascomal wall brown, composed of flattened cells. Asci unitunicate, ellipsoid, with apical ring. Ascospores hyaline, 2-celled, with septum near base, ellipsoid to ovoid or straight, with rounded ends.

Anamorph: *Cylindrosporella.*

Habitat: In leaves of deciduous trees.

Representative species: *Stegophora ulmea* (Schwein.:Fr.) Syd. & P. Syd. [Anam. *Cylindrosporella ulmea* (Miles) Arx], cause of leafspot of elm *(Ulmus* spp). This fungus has also been known as *Gnomonia ulmea* (Schwein.:Fr.) Thuem., *Dothidella ulmea* (Schwein.:Fr.) Ellis & Everh., and *Lambro ulmea* (Schwein.:Fr.) E. Müller.

Comments: In *Sphaerognomonia* Potebnia the ascospores are one-celled; in *Linocarpon* Syd. & P. Syd. the ascospores are septate and usually filiform.

Reference: Barr, 1978.

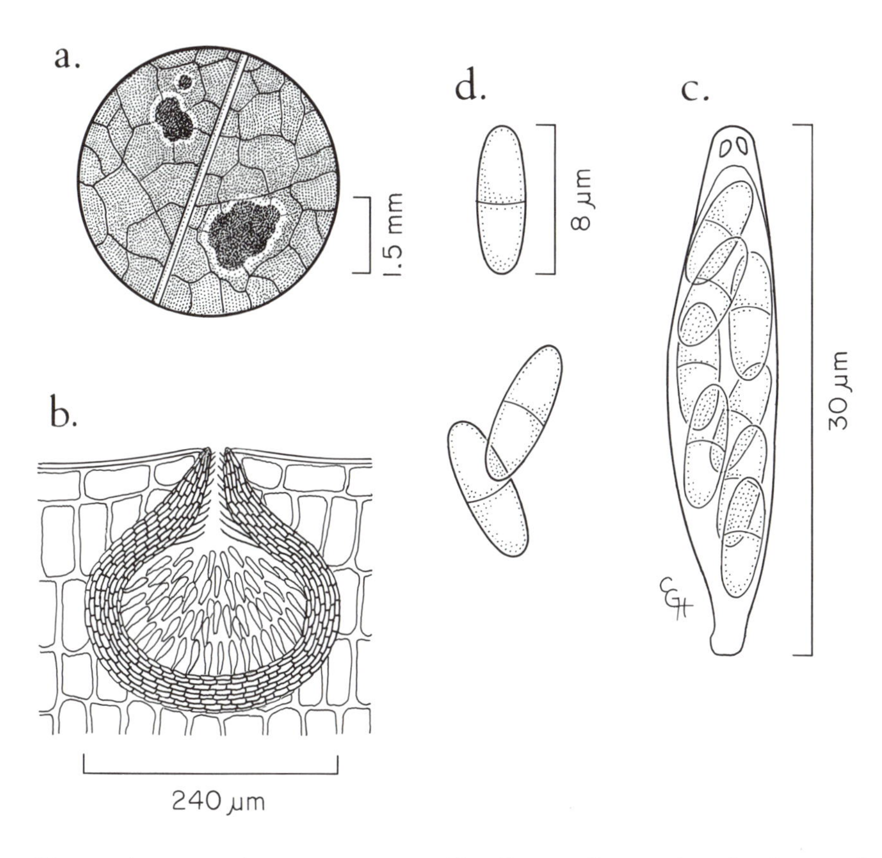

Stegophora ulmea. **a.** Section through leaf and perithecium with asci. **b.** Portion of elm leaf with lesions. **c.** Ascus with ascospores. **d.** Mature ascospores.

NECTRIA (Fr.) Fr.

Ascoma an ostiolate perithecium, solitary to gregarious, with or without a stroma, if stromatic, perithecium superficial or partially immersed in stroma, fleshy, brightly-colored to occasionally light brown, subglobose to globose or obpyriform. Ascomal wall pseudoparenchymatous or prosenchymatous, walls of outermost cells pigmented and somewhat thickened, inner cells hyaline and thin-walled. Centrum containing apical paraphyses. Asci unitunicate, clavate to cylindrical, with persistent wall, with or without an apical ring, nonamyloid, straight to curved, 8-spored. Ascospores 2-several-celled, ellipsoid to long fusiform, hyaline to brownish, smooth, striate or verrucose. (Fig. A-B,D).

Anamorphs: *Acremonium, Antipodium, Botryocrea, Calostilbella, Cephalosporiopsis, Ciliciopodium, Cylindrocarpon, Cylindrocladiella, Cylindrocladium, Dendrodochium, Didymostilbe, Fusarium, Gliocladium, Kutilakesopsis, Myriothecium, Rhizostilbella, Stilbella, Tubercularia, Verticillium, Virgatospora, Volutella, Zythia,* and *Zythiostroma.*

Habitat: Occurring on a wide range of woody and herbaceous plants and cryptogams as saprobes or parasites, and in soil.

Representative species: *Nectria cinnabarina* (Tode:Fr.) Fr., (Anam. *Tubercularia vulgaris* Tode:Fr.), a stromatic species that causes canker diseases of trees and shrubs, and *N. haematococca* Berk. & Broome [Anam. *Fusarium solani* (Mart.) Sacc.], a nonstromatic species that causes root diseases of crop plants.

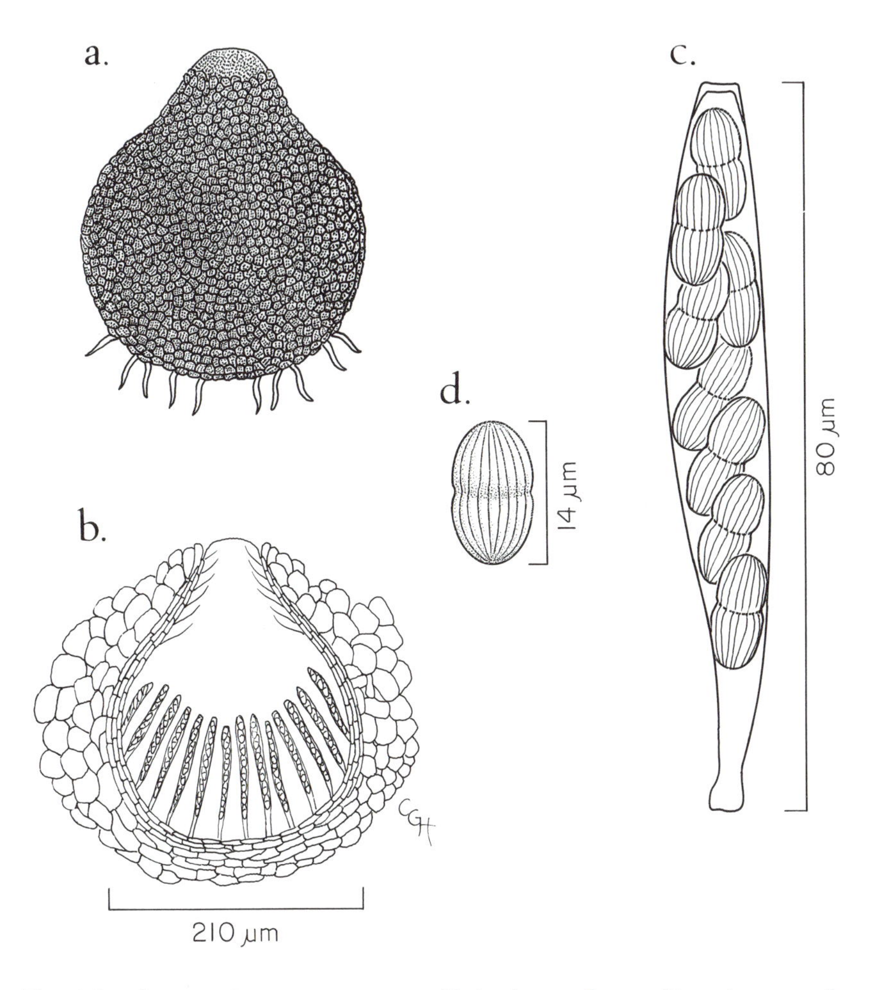

Nectria haematococca. **a.** Exterior of perithecium. **b.** Section through perithecium with asci. **c.** Ascus with ascospores. **d.** Mature ascospore with striations.

Comments: Traditionally *Nectria* has been used for perithecial fungi with brightly-colored ascomata and 2-celled, hyaline ascospores, but Rossman includes species with phragmosporous ascospores as well, on the basis of ascomal wall structure and the anamorph. Some authors have restricted *Nectria* to non-stromatic species, placing species with a stroma in *Creonectria* Seaver or *Dialonectria* (Sacc.) Cooke. *Sphaerostilbe* Tul. has been used for species with a stilboid anamorph, and *Neuronectria* Munk has been used for species with striate ascospores in which the ascomata collapse and become cupulate following ascospore discharge. All are included in *Nectria.*

References: Booth, 1959; Dingley, 1951; Domsch et al., 1980; Ellis and Ellis, 1985; Ellis and Everhart, 1892; Matsushima, 1971, 1975; Rossman, 1983; Samuels, 1976; Samuels and Dumont, 1982; Viégas, 1944; BK 324-331; CMI 21, 147-148, 391, 531-533, 623-624, 715; IM 10:C142.

➔

Fig. 4. **A-B.** *Nectria haematococca*. **A.** Exterior of perithecium. X212. **B.** Ascus with ascospores. X955. **C.** *Botryosphaeria dothidea* ascus with ascospores. X500. **D.** Mature ascospores of *Nectria haematococca*. X957. **E.** Section through erumpent ascostroma of *Botryosphaeria dothidea*. X65. **F-H.** *Myriogenospora atramentosa*. **F.** Section through perithecium on grass leaf. X167. **G.** Part-spores. X281. **H.** Asci filled with part-spores. X277.

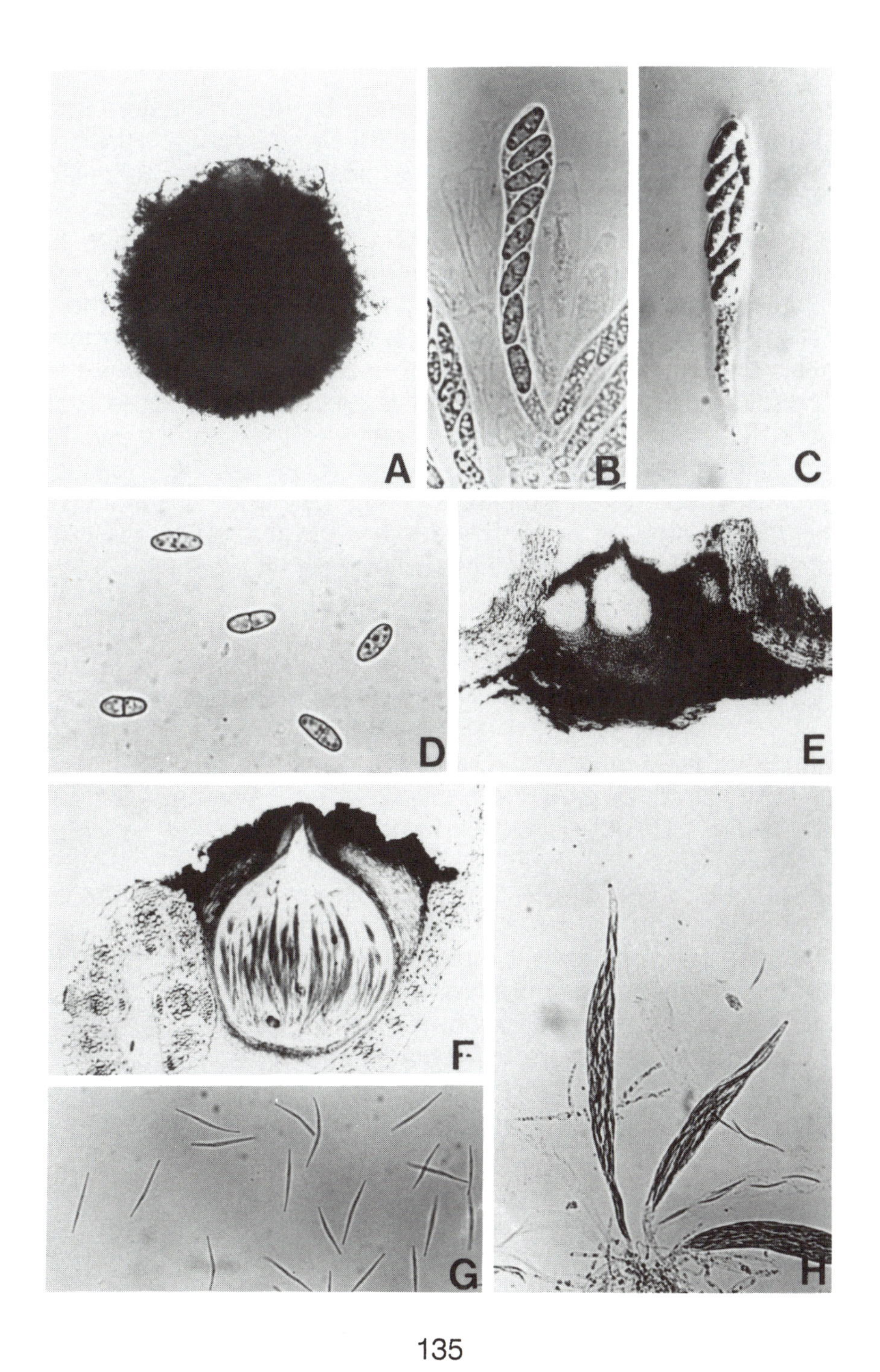
A
B
C
D
E
F
G
H

DIAPORTHE Nitschke

Ascoma an ostiolate perithecium, formed in clusters in a stroma composed of both host and fungal tissues (pseudostroma), with the outer margin of the stroma delimited by a broad blackened zone. Perithecia in groups or sometimes separate, subglobose to obpyriform, oblique to lateral, with ostiolar necks converging and erumpent through outer tissues of host. Ascomal wall two-layered, the outermost cells isodiametric, thick-walled and black, the inner cells hyaline, thin-walled and flattened. Asci unitunicate, clavate to ellipsoid, with nonamyloid, refractive apical ring, 8-spored, wall persistent, but with deliquescent base so that mature asci lie free in perithecium at maturity. Ascospores hyaline, 2-celled, septum median, constricted or not, cylindric to fusoid or cylindrical, straight or slightly curved and inaequilateral, with short, narrow appendages in some species.

Anamorph: *Phomopsis.*

Habitat: On stems of woody and herbaceous plants.

Representative species: *Diaporthe phaseolorum* (Cooke & Ellis) Sacc. [Anam. *Phomopsis phaseoli* (Desmaz.) Sacc.], cause of blight of soybean and other plants.

Comments: *Diaporthopsis* Fabre differs from *Diaporthe* in having 1-celled ascospores. *Apioporthe* Höhn. has been used for species with apiosporous ascospores, but it is now considered synonymous with *Anisogramma* Theiss. & Syd. In *Cryptodiaporthe* Petr. the entostroma is lacking or sparse and there is no blackened zone.

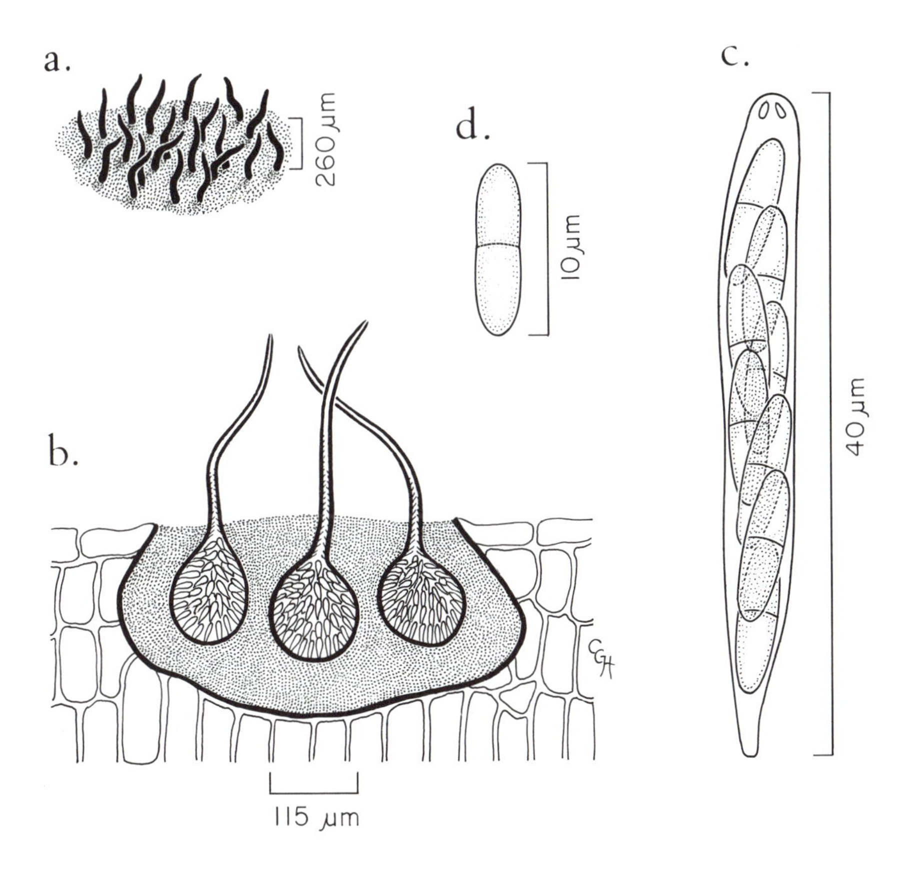

Diaporthe phaseolorum. **a.** Ostiolar necks protruding from stroma in host. **b.** Section through stroma with perithecia. **c.** Ascus with ascospores. **d.** Mature ascospore.

References: Dennis, 1981; Ellis and Ellis, 1985; Ellis and Everhart, 1892; Kobayashi, 1970; Munk, 1957; Wehmeyer, 1933; BK 363-366; CMI 336, 396, 733-734; FC 70; IM 19:C271.

➛

Fig. 5. **A-C.** *Tuber texense*. **A.** Exterior and interior views of ascoma. X0.6. **B.** Mature ascospore. X850. **C.** Asci with ascospores. X225. **D-E.** *Microascus cirrosus*. **D.** Exterior of ascoma. X187. **E.** Mature ascospores. X844. **F-G.** *Neocosmospora tenuicristata*. **F.** Exterior of perithecium. X233. **G.** Section through perithecium with asci. X267. **H-J.** *Balansia henningsiana*. **H.** Ascus with ascospores. X143. **I.** Perithecial stroma on grass leaf. X31. **J.** Mature ascospores. X285.

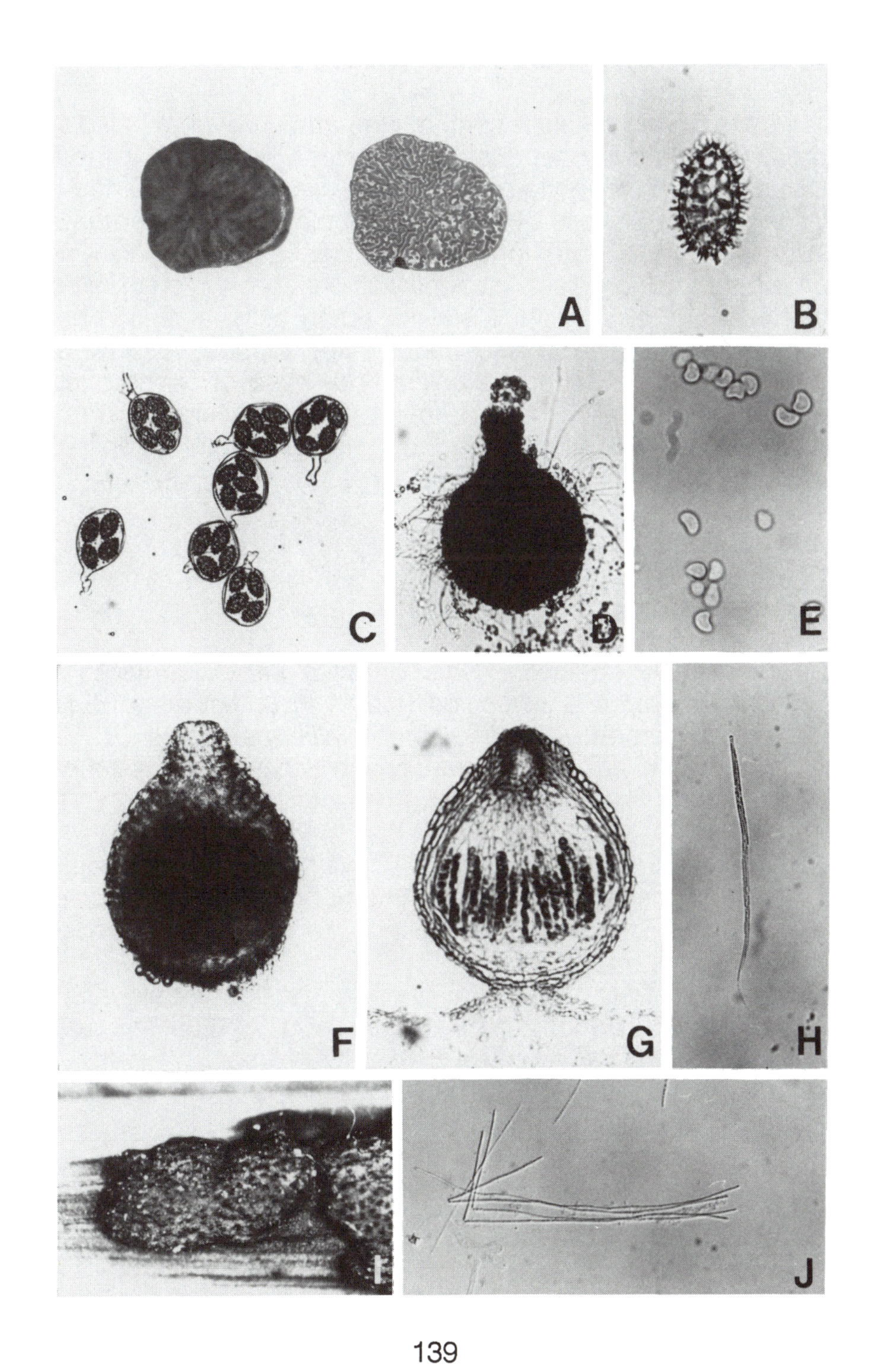
A
B
C
D
E
F
G
H
I
J

CRYPHONECTRIA (Sacc.) Sacc.

Ascoma an ostiolate perithecium, immersed in a valsoid, prosenchymatous pseudostroma; stroma immersed in host tissues, usually brightly-colored, with distinct disc erumpent through host tissues. Perithecia subglobose, clustered in stroma, oblique, with long ostiolar necks converging and erumpent through disc. Ascomal wall dark brown, composed of flattened, slightly thick-walled outer cells and hyaline, thin-walled inner cells. Asci unitunicate, ellipsoid to clavate, with nonamyloid, refractive apical ring, base of ascus often delinquescent, asci lying free at maturity, eight-spored. Ascospores hyaline, two-celled, with median septum, ellipsoid or ovoid.

Anamorph: *Endothiella.*

Habitat: On branches of deciduous trees.

Representative species: *Cryphonectria parasitica* (Murrill) Barr [=*Endothia parasitica* (Murr.) P. J. Anderson & H. W. Anderson] (Anam. *Endothiella parasitica* P. J. Anderson), cause of blight disease of chestnut (*Castanea dentata*).

Comments: *Endothia* Fr. differs in having a diatrypoid stroma and 1-celled ascospores. In *Amphiporthe* Petr. the stroma is dark-colored and the ascospores are narrowly ellipsoid.

References: Barr, 1978; Dennis, 1981; Hodges et al., 1986; Micales and Stipes, 1987; Roane et al., 1986; CMI 449, 704; IM 9:C126.

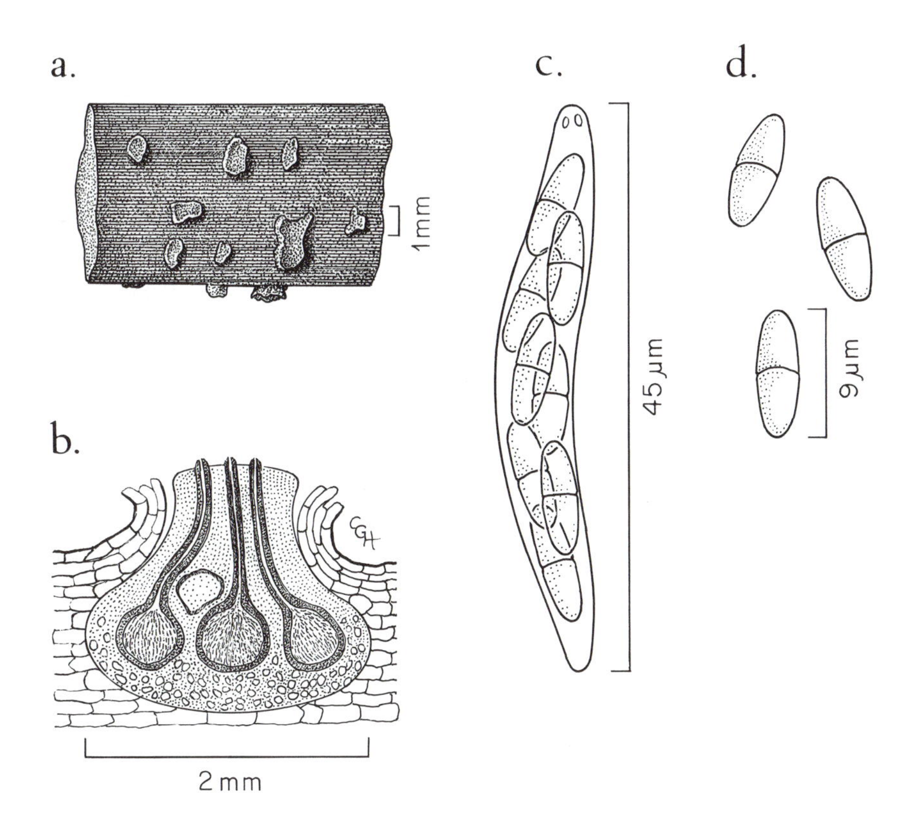

Cryphonectria parasitica. **a.** Erumpent stromata on chestnut branch. **b.** Section through stroma with perithecia and remains of pycnidium. **c.** Ascus with ascospores. **d.** Mature ascospores.

HYPOCREA Fr.

Stromata formed superficially on surface of substrate, pulvinate to effuse or irregular in outline, exterior brightly colored or black, interior hyaline, fleshy. Ascoma an ostiolate perithecium, subglobose to ovate, immersed in stroma with slightly protruding ostiolar papilla. Ascomal wall composed of flattened, elongate cells. Centrum containing apical paraphyses. Asci unitunicate, cylindrical, with or without apical apparatus, nonamyloid, 8-spored, but sometimes appearing 16-celled. Ascospores 2-celled, cells rounded and deeply constricted at septum, often separating into subglobose part-spores, hyaline or greenish. (Fig. 6A-C).

Anamorph: *Acremonium, Gliocladium, Trichoderma,* and *Verticillium*.

Habitat: Occurring on dead wood.

Representative species: *Hypocrea citrina* Pers.:Fr.) Fr., with lemon yellow stromata.

Comments: Species of *Hypocrea* with green, olivaceous or brownish spores have been placed in *Chromocrea* Seaver and in *Creopus* Link. Species with a stilboid anamorph are placed in *Stilbocrea* Pat.

References: Dingley, 1952a; Doi, 1966, 1972; Ellis and Ellis, 1985; Ellis and Everhart, 1892; Seaver, 1910; BK 317-321; IM 23:C353.

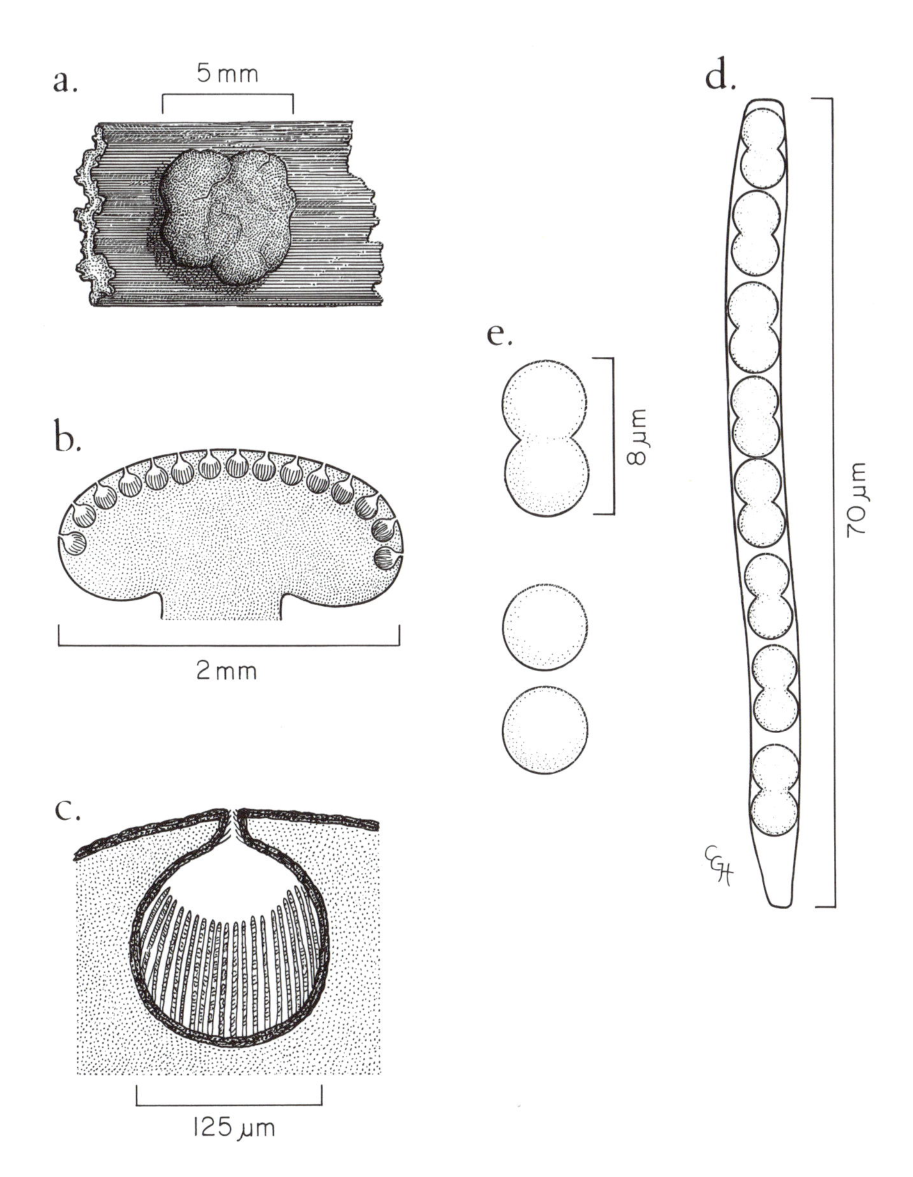

Hypocrea schweinitzii. **a.** Perithecial stroma on wood. **b.** Section through stroma with perithecia. **c.** Section through perithecium with asci. **d.** Mature ascus. **e.** Above, mature ascospore; below, part-spores.

HYPOMYCES (Fr.) Tul.

Mycelium forming an effuse, often cottony subiculum on surface of substrate, brightly colored, somewhat fleshy; ascoma an ostiolate perithecium, immersed in subiculum, with protruding ostiolar necks. Ascomal wall thin, composed of elongated cells, slightly darker on exterior. Centrum containing apical paraphyses. Asci unitunicate, cylindrical, with thick, nonamyloid apical cap, 8-spored. Ascospores hyaline, 1-2-celled, fusoid to fusiform, with an apiculus at each end, usually roughened. (Fig. 6D-G).

Anamorphs: *Clathrobotryum, Mycogone, Sepedonium, Sibirina, and Trichothecium*.

Habitat: On wood, stems or other fungi.

Representative species: *Hypomyces lactifluorum* (Schwein.) Tul., parasitic on russulaceous agarics.

Comments: The genus *Apiocrea* Syd. has been used for 2-celled species with a nonmedian septum, *Arachnocrea* Moravec has been used for 2-celled species in which the spores separate into part-spores, and species with 1-celled ascospores have been segregated into *Peckiella* (Sacc.) Sacc.

References: Dennis, 1981; Ellis and Everhart, 1892; Müller and von Arx, 1962; Munk, 1957; Rogerson and Samuels, 1985; Seaver, 1910; Tubaki, 1960, 1975; BK 322-323; IM 29:C528.

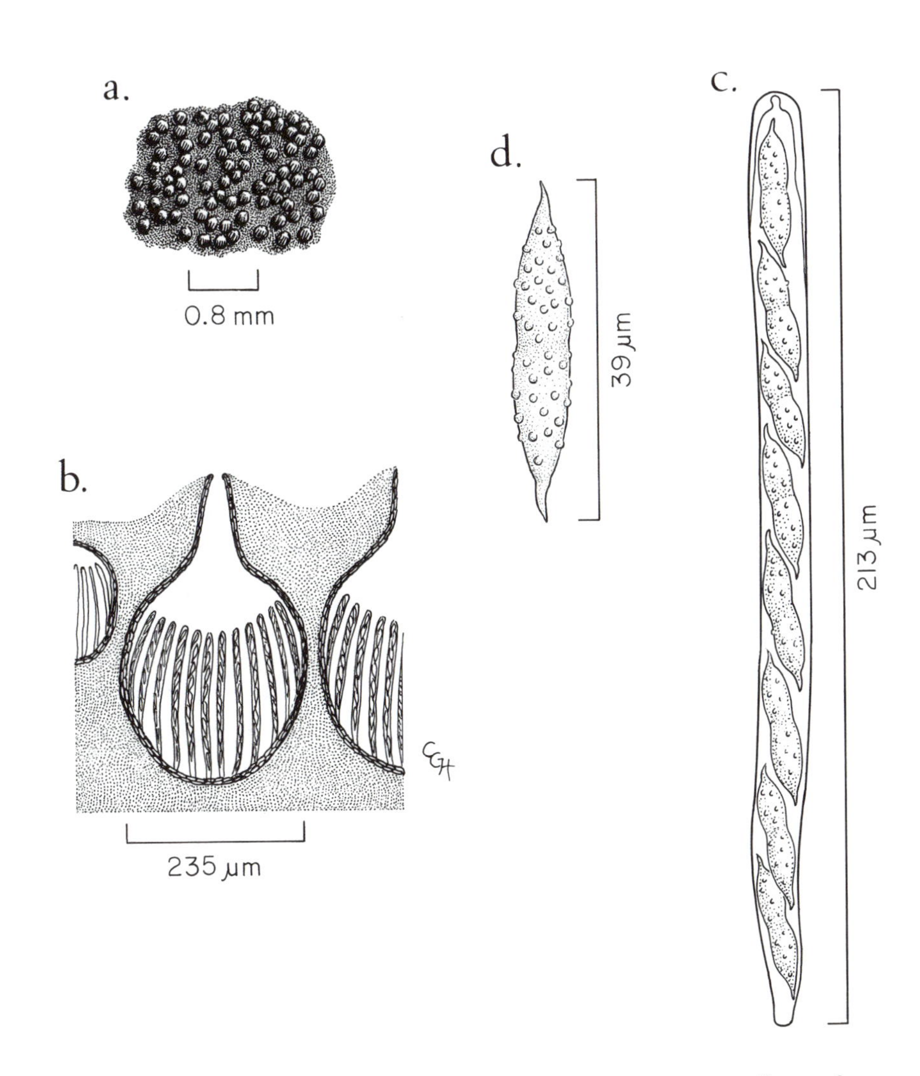

Hypomyces lactifluorum. **a.** Ostioles protruding from stroma on mushroom. **b.** Section through perithecium with asci. **c.** Ascus with ascospores. **d.** Mature ascospore.

DIDYMELLA Sacc.

Ascoma a uniloculate perithecioid pseudothecium, scattered, immersed in substrate or erumpent; pseudothecia dark brown to black, globose to subglobose, ostiolate. Ascomal wall pseudo- parenchymatous, sometimes thickened around ostiole to form a clypeus. Centrum containing filamentous, often branched, pseudoparaphyses. Asci bitunicate, parallel, clavate to cylindrical, short-stalked, 8-spored. Ascospores 2-celled, septum median, constricted or not, oval to fusiform, hyaline.

Anamorphs: *Ascochyta, Phloeospora, Phoma*, and *Stagonospora.*

Habitat: In leaves and stems of angiosperms and conifers.

Representative species: *Didymella bryoniae* (Auersw.) Rehm (Anam. *Ascochyta cucumis* Fautr. & Roum.), cause of gummy stem blight of cucurbits. This species has also been known as *Didymella melonis* Pass., *Mycosphaerella citrullina* (C. O. Sm.) Gross., *M. cucumis* (Fautr. & Roum.) Chiu & J. C. Walker, and *M. melonis* (Pass.) Chiu & J. C. Walker.

Comments: This genus is often confused with *Mycosphaerella* Johans., from which it differs in having ascospores that are broader and more constricted, the parallel arrangement of the asci, the presence of pseudoparaphyses, larger ascomata, and in having different anamorphs.

References: Corbaz, 1956; Corlett, 1981; Dennis, 1981; Ellis and Ellis, 1985; Holm, 1953; Müller and von Arx, 1962; Munk, 1957; Sivanesan, 1984; CMI 272, 332, 622, 633, 735; FC 49, 131, 303; IM 20:C296.

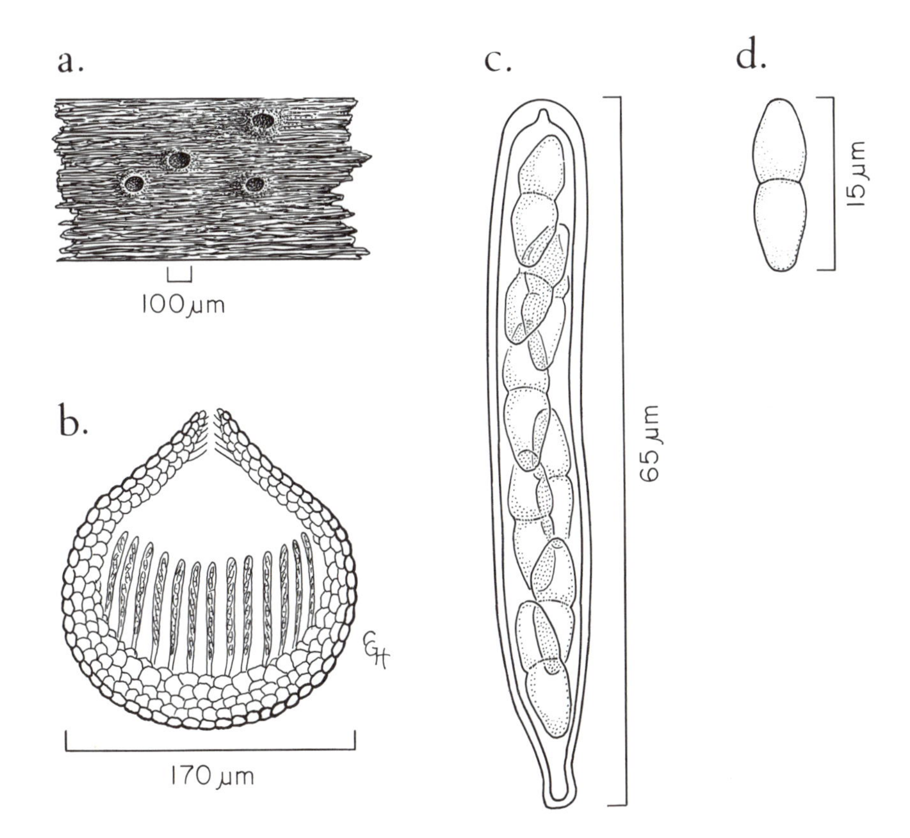

Didymella bryoniae. **a.** Erumpent pseudothecia on watermelon stem. **b.** Section through pseudothecium with asci. **c.** Ascus with ascospores. **d.** Mature ascospore.

MYCOSPHAERELLA **Johans.**

Ascoma a uniloculate, perithecioid pseudo- thecium, erumpent or immersed in host tissues, single to sometimes aggregated; pseudothecia very small, dark brown or usually black, globose, conical or depressed. Ascomal wall 1-4-layered, composed of dark brown polygonal cells. Centrum pseudoparenchymatous. Asci bitunicate, oblong, elongate, ovoid or saccate to occasionally clavate, often ventricose, formed in fascicles, 8-spored. Ascospores, 2-celled, septum median or nearly so, constricted or not, hyaline, but sometimes pale brown in age, variable in shape, fusoid, oblong, obovate or elongate, smooth, but occasionally roughened with age, sometimes with a thin mucilaginous covering.

Anamorphs: Cercoseptoria, *Cercospora, Cercosporella, Cercosporidium, Cladosporium, Colletogloeum, Fusicladiella, Miuraea, Ovularia, Paracercospora, Phaeoisariopsis, Phloeospora, Pseudocercospora, Pseudocercosporella, Ramularia, Septoria, Stenella*, and *Stigmina.* Microconidial and spermatial states are also found in some species.

Habitat: Mostly in leaves and stems of herbaceous plants, and in leaves of deciduous trees, often parasitic.

Representative species: *Mycosphaerella arachidis* Deighton (Anam. *Cercospora arachidicola* S. Hori) and *M. berkeleyi* W. A. Jenkins [Anam. *Cercosporidium personatum* (Berk. & M. A. Curtis) Deighton =*Phaeoisariopsis personata* (Berk. & M. A. Curtis) Arx], cause of early and late leafspot diseases of peanut *(Arachis* spp.).

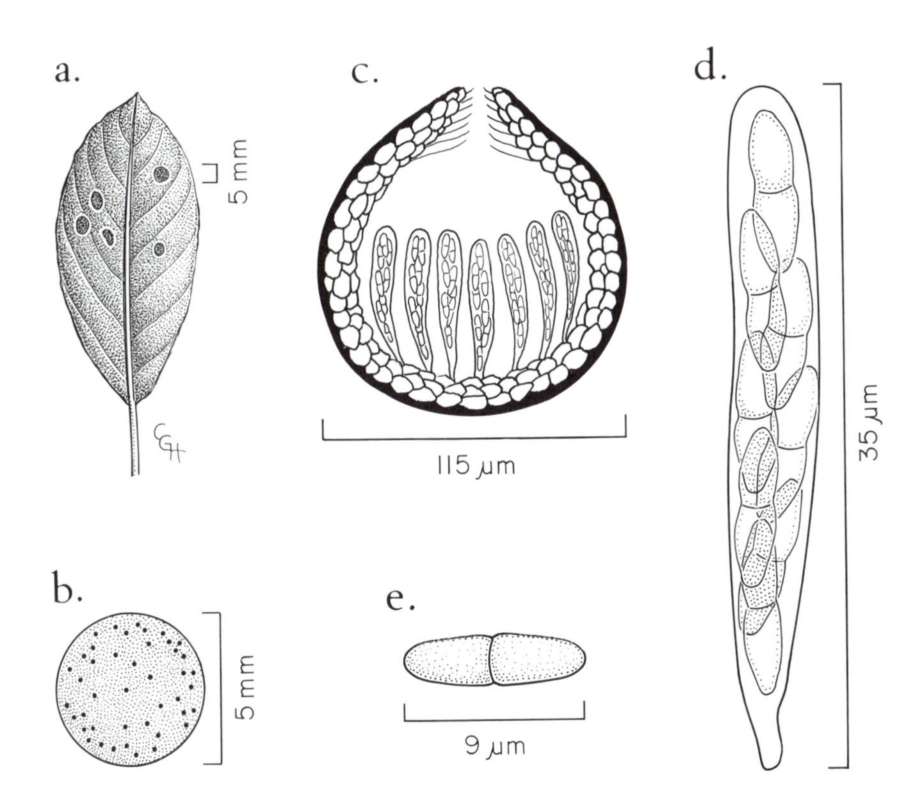

Mycosphaerella berkeleyi. **a.** Lesions on peanut leaf. **b.** Close-up of lesion with dark ostiolar necks. **c.** Section through pseudothecium with asci. **d.** Ascus with ascospores. **e.** Mature ascospore.

Comments: In *Delphinella* (Sacc.) Kuntze the asci are polysporous, and the pseudothecia are larger, erumpent, and sometimes laterally expanded.

References: Arx, 1949; Barr, 1972; Dennis, 1981; Ellis and Ellis, 1985; Müller and von Arx, 1962; Munk, 1957; Sivanesan, 1984; BK 376; CMI 340, 411-414; 435, 468, 510, 584, 708-709; IM 6:C90.

→

Fig. 6. **A-C.** *Hypocrea schweinitzii*. **A.** Section through stroma with peripheral row of perithecia. X30. **B.** Separated cells of ascospore (part-spores). X2520. **C.** Close-up of perithecium with asci. X251. **D-G.** *Hypomyces lactifluorum*. **D.** Close-up of perithecium with asci. X133. **E.** Ascus with ascospores. X239. **F.** Perithecia immersed in stroma. X63. **G.** Mature apiculate ascospore. X951. **H-I.** *Lembosia melastomum*. **H.** Longitudinal section through ascostroma with asci. X686. **I.** Mature ascospores. X462.

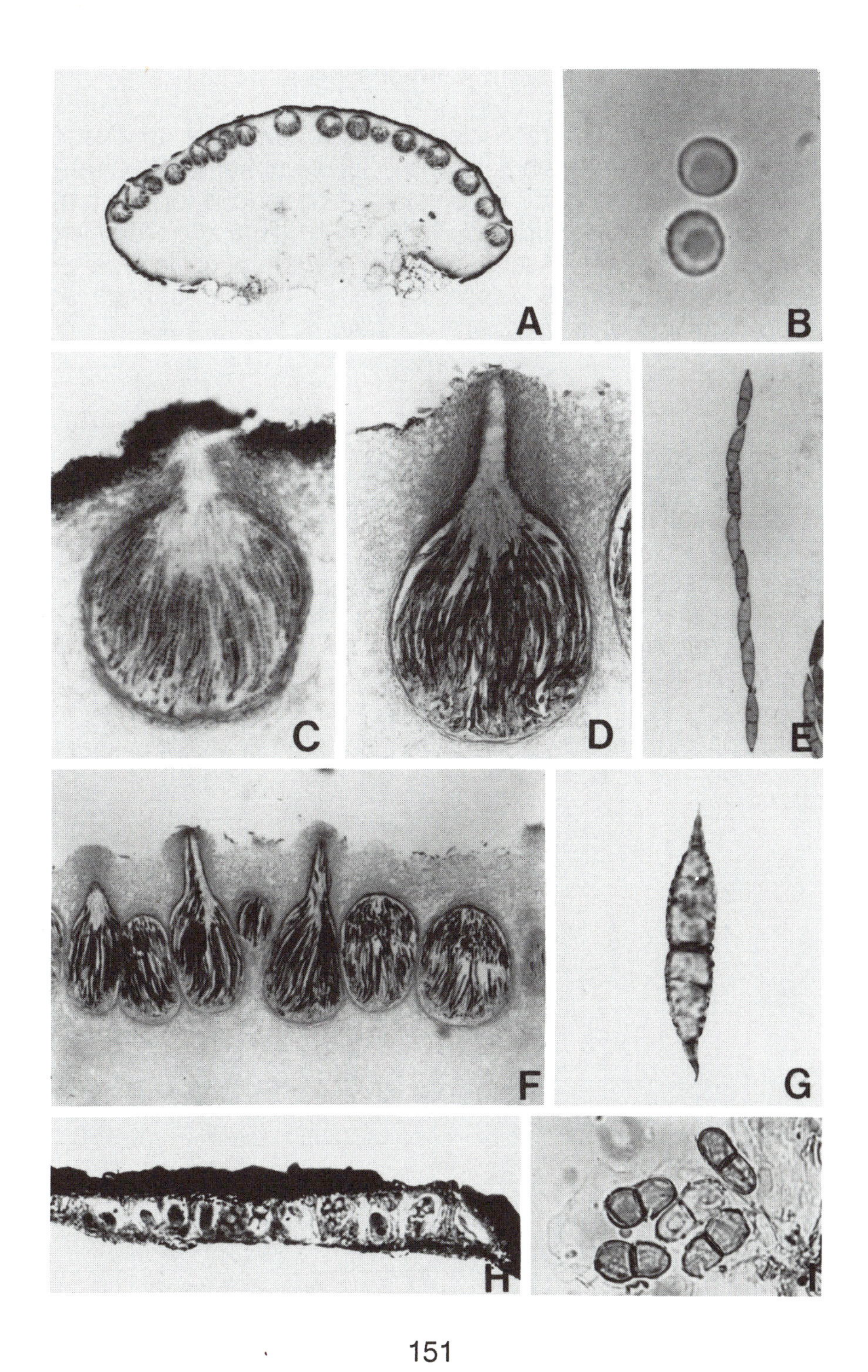
A
B
C
D
E
F
G
H
I

GLONIUM Mühlenberg

Ascoma a hysterothecium, oval to elongate or linear, branched or stellate, single, superficial or sunken in substrate, with or without a subiculum, black, carbonaceous, opening by a sunken longitudinal slit. Centrum containing pseudoparaphyses. Asci bitunicate, clavate to cylindrical or saccate, 8-spored. Ascospores 2-celled, hyaline, sometimes becoming light yellow-brown with age.

Anamorph: *Hysteropycnis.*

Habitat: On wood and bark of trees and shrubs.

Representative species: *Glonium stellatum* Mühlenberg.

Comments: In Farlowiella Sacc. the ascospores are 1-2-celled and brown. *Gloniella* Sacc. and *Gloniopsis* De Not. have hyaline ascospores; they are phragmosporous in *Gloniella* and dictyosporous in *Gloniopsis.*

References: Dennis, 1981; Müller and von Arx, 1962; Zogg, 1962; IM 18:C254.

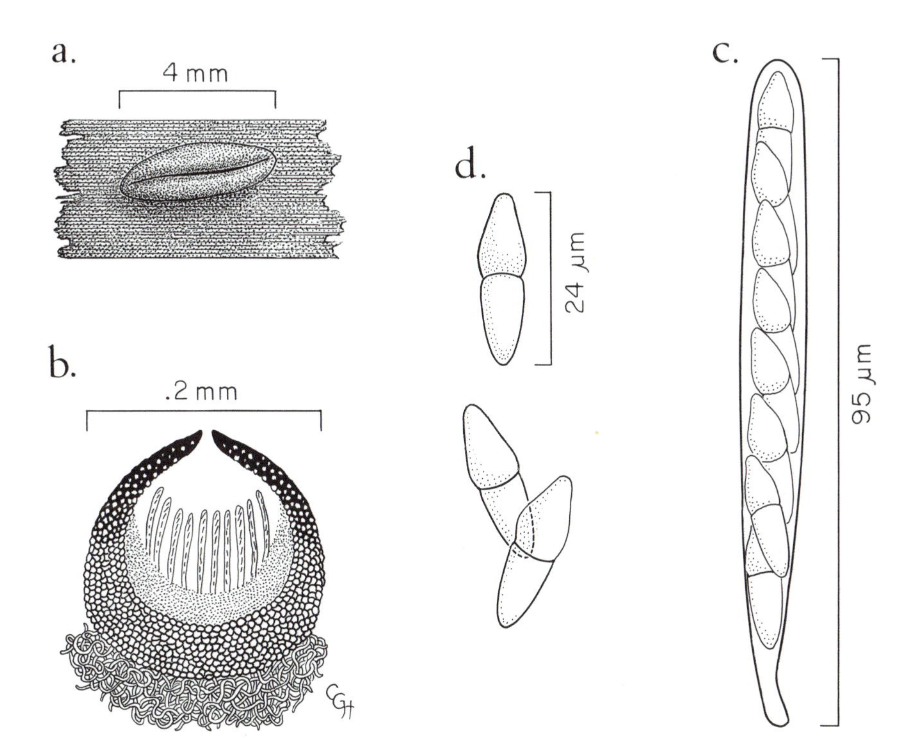

Glonium stellatum. **a.** Hysterothecium on wood. **b.** Section through hysterothecium with asci. **c.** Ascus with ascospores. **d.** Mature ascospores.

SCHIZOTHYRIUM **Desmaz.**

Ascoma a dark, dimidiate pseudothecium formed superficially on living plant tissues, appressed to cuticle; pseudothecium brown or black, seated on a thin, hyaline basal stroma and covered by a dark hyphal wall that splits open at maturity. Centrum containing slender filaments. Asci bitunicate, formed in a single layer, clavate or subglobose, 8-spored. Ascospores 2-celled, hyaline, ellipsoidal.

Anamorphs: *Leptothyrium* and *Zygophiala.*

Habitat: On living leaves, stems and fruits of flowering plants.

Representative species: *Schizothyrium pomi* (Mont. & Fr.) Arx, (Anam. *Zygophiala jamaicensis* E. Mason) cause of flyspeck disease of apples and pears.

Comments: In *Leptophyma* Sacc. the ascoma is bright-colored to subhyaline and gelatinous.

References: Ellis and Ellis, 1985; Müller and von Arx, 1962.

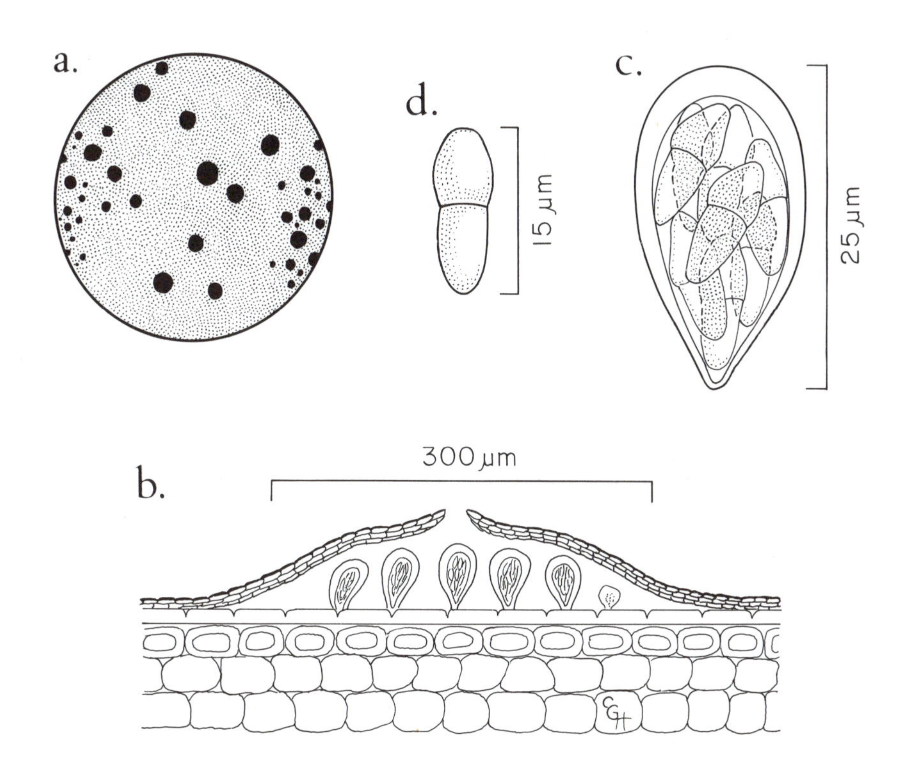

Schizothyrium pomi. **a.** Ascomata on apple. **b.** Section through ascoma with asci. **c.** Ascus with ascospores. **d.** Mature ascospore.

BOMBARDIA Fr.

Ascoma an ostiolate perithecium, narrowly ellipsoidal to clavate, lacking a neck, aggregated, brown to black, tapering to an elongated base in the interior; ascomal wall 5-layered (bombardioid), the two outermost layers regarded as stromatic; the outer layer thin and dark, with an indistinct structure, the middle layer thick, hyaline to subhyaline, cartilaginous, traversed by slender, branched hyphae, merging into a thin layer of fibrous hyphae; the two innermost layers regarded as the true perithecial wall, the outermost of these layers composed of dark-brown, angular cells that are black and carbonaceous toward the base, merging into the innermost layer of flattened, thin-walled cells. Paraphyses abundant, filiform, branched, mixed with the asci. Asci unitunicate, clavate, with a truncate apex and a simple, nonamyloid apical ring. Ascospores cylindrical and sigmoid when formed, with numerous oil droplets, caudate at both ends, becoming septate, with the upper cell swelling and becoming dark brown, ellipsoidal to ovoid, with a truncate base and an upper germ pore; lower cell remaining hyaline, cylindrical or bent, finally collapsing.

Anamorph: None reported.

Habitat: On dead wood.

Representative species: *Bombardia bombarda* (Fr.) J. Schröt.

Comments: *Bombardia* is unusual in the vertically elongated ascoma surrounded by a thin stroma, which distinguishes it from *Cercophora Fuckel,* in which the stroma is lacking. Like *Cercophora,* the ascospores are capable of germinating while still hyaline.

References: Cailleux, 1971; Dennis, 1981; Ellis and Ellis, 1985; Ellis and Everhart, 1892; Lundqvist, 1972; Munk, 1957; IM 9:C125.

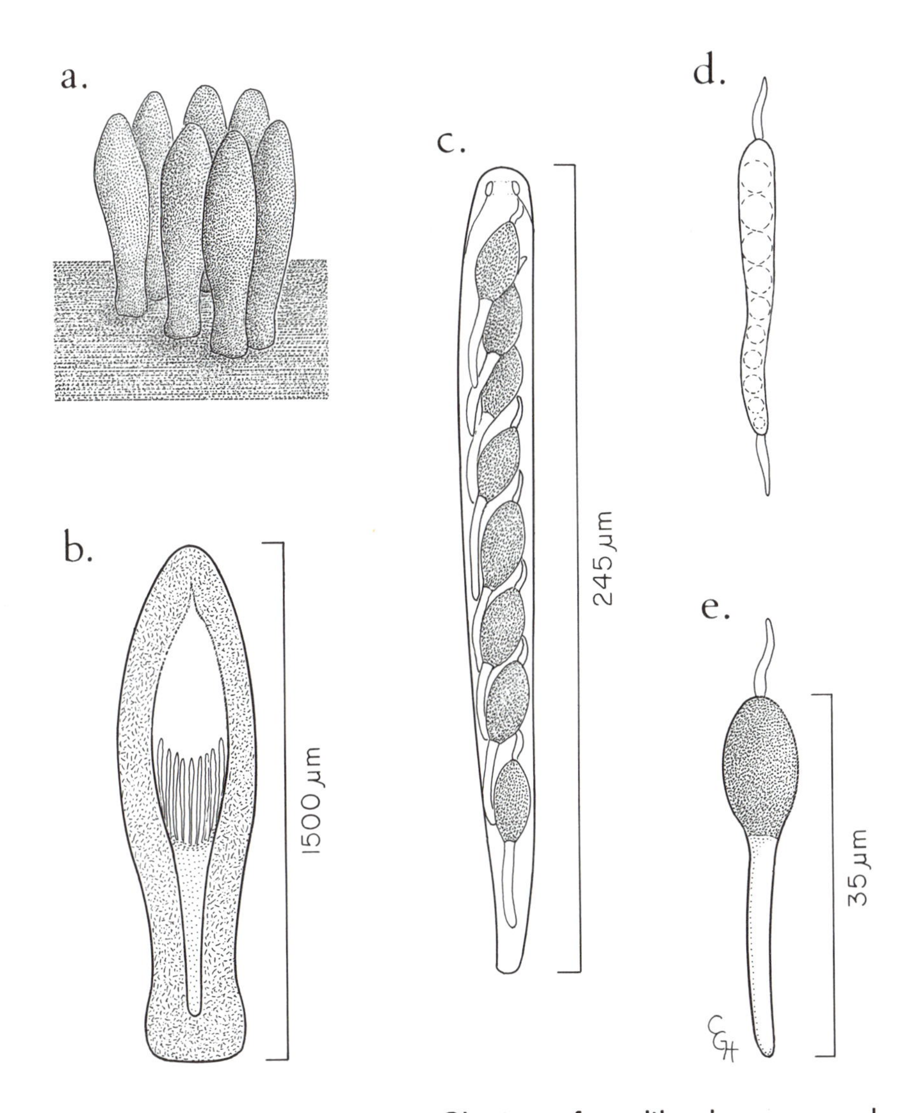

Bombardia bombarda. **a.** Cluster of perithecia on wood. **b.** Section through perithecium. **c.** Ascus with ascospores. **d.** Young ascospore with terminal gelatinous appendages. **e.** Mature ascospore with gelatinous appendage on upper end.

CERCOPHORA Fuckel

Ascoma an ostiolate perithecium, obpyriform to conical or subglobose, dark-brown to black, usually covered with septate hairs or setae; wall 3-4-layered, pseudoparenchymatous or pseudobombardioid, with cells of outer layer angular, dark, with somewhat thickened walls, cells of middle layer hyaline to subhyaline with thickened walls, often traversed by slender, branched hyphae, merging into an inner layer of flattened, thin-walled, hyaline cells. Paraphyses filamentous to ventricose, mixed with the asci. Asci unitunicate, clavate, with nonamyloid apical ring, 8-spored, with spores 1-3-seriate. Ascospores at first cylindrical and straight, caudate at both ends, becoming sigmoid and septate, upper cell swelling and becoming brown, ellipsoidal to ovoid, truncate at base, with upper germ pore, smooth, sometimes becoming septate; lower cell remaining hyaline, sometimes becoming septate, collapsing at maturity. Spores can be discharged at the hyaline stage and will germinate.

Anamorphs: *Cladorrhinum* and *Phialophora.*

Habitat: On dung, wood, or in soil.

Representative species: *Cercophora mirabilis* Fuckel, on dung.

Comments: This genus differs from *Podospora* Ces. in having cylindrical spore initials that become sigmoid, in the tendency for the spores to discharge and germinate while still hyaline, and in the usually pseudobombardioid wall. It differs from *Bombardia* Fr. in the lack of a stroma surrounding the perithecial wall.

References: Cailleux, 1971; Dennis, 1981; Ellis and Ellis, 1985; Hanlin and Tortolero, 1987; Lundqvist, 1972; Munk, 1957; IM 9:C125.

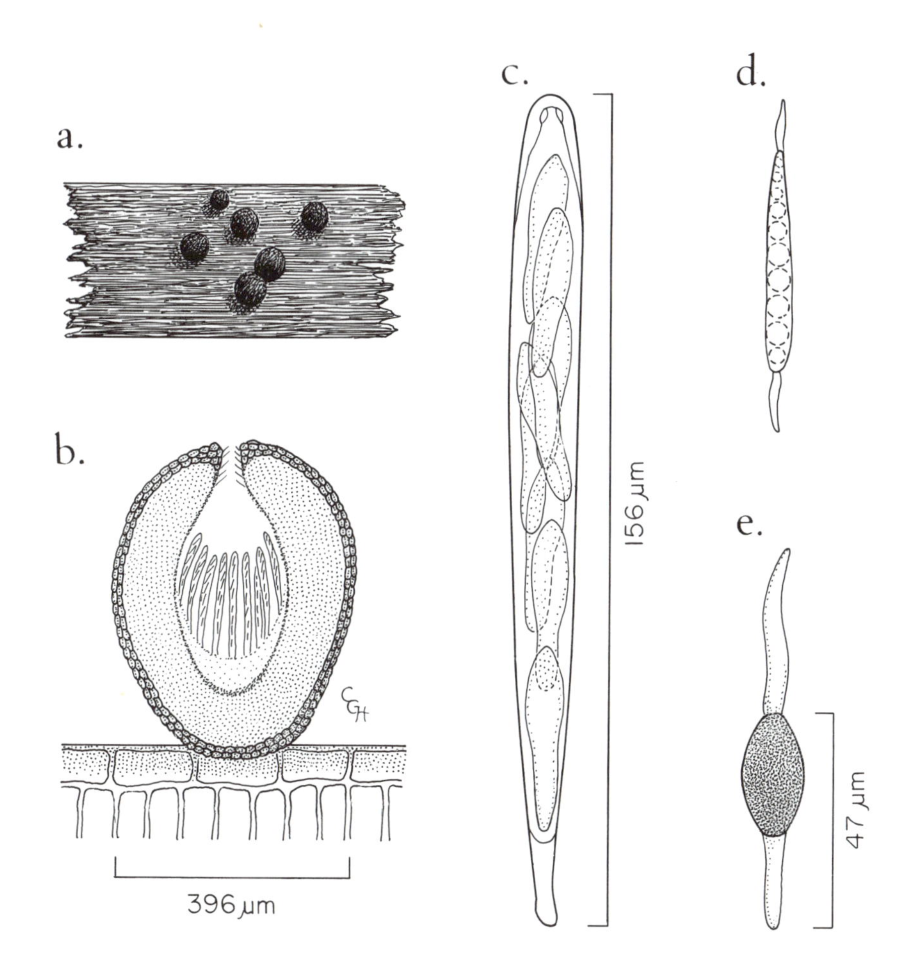

Cercophora palmicola. **a.** Perithecia on palm stem. **b.** Section through perithecium with asci. **c.** Ascus with immature ascospores. **d.** Young ascospore with terminal gelatinous appendages. **e.** Mature ascospore with gelatinous appendage on upper end.

PODOSPORA Ces.

Ascoma an ostiolate, obpyriform perithecium, nonstromatic, often covered with hairs of different types. Ascomal wall 3-layered and membranaceous or 4-layered and pseudobombardioid and coriaceous; in 3-layered wall, cells of outer layer angular and dark, cells of middle layer flattened and hyaline, merging into the inner layer of flattened, thin-walled, hyaline cells; in pseudobombardioid wall, cells of outer layer dark and angular, with slightly thickened walls, second layer thick and composed of subhyaline cells with very thick walls, and an inner layer of flattened, hyaline, thin-walled cells. Paraphyses filamentous to ventricose, simple, mixed with the asci. Asci unitunicate, clavate to saccate, rarely cylindrical, with a thin, nonamyloid apical ring, but ring sometimes lacking, 4-multi-spored. Ascospores clavate when delimited, or cylindrical and soon becoming clavate, with one to several gelatinous appendages at each end, swelling above and becoming septate, upper cell ellipsoidal, becoming dark brown to blackish, with one or several germ pores near apex, truncate at base, smooth; lower cell remaining hyaline and eventually collapsing.

Anamorph: None, but microconidia are formed in some species.

Habitat: Mainly coprophilous, sometimes on seeds and other plant parts.

Representative species: *Podospora fimeseda* (Ces. & De Not.) Niessl.

Comments: The generic name *Pleurage* Fr. has been applied to these fungi by various authors and is a synonym of *Podospora. Malinvernia* Rabenh. was used for a species with 4-spored asci, and *Philocopra* Speg. has been used for species with multispored asci; both genera are here included in *Podospora. Andreanszkya* Tóth was erected for a species in which the primary appendage (pedicel)

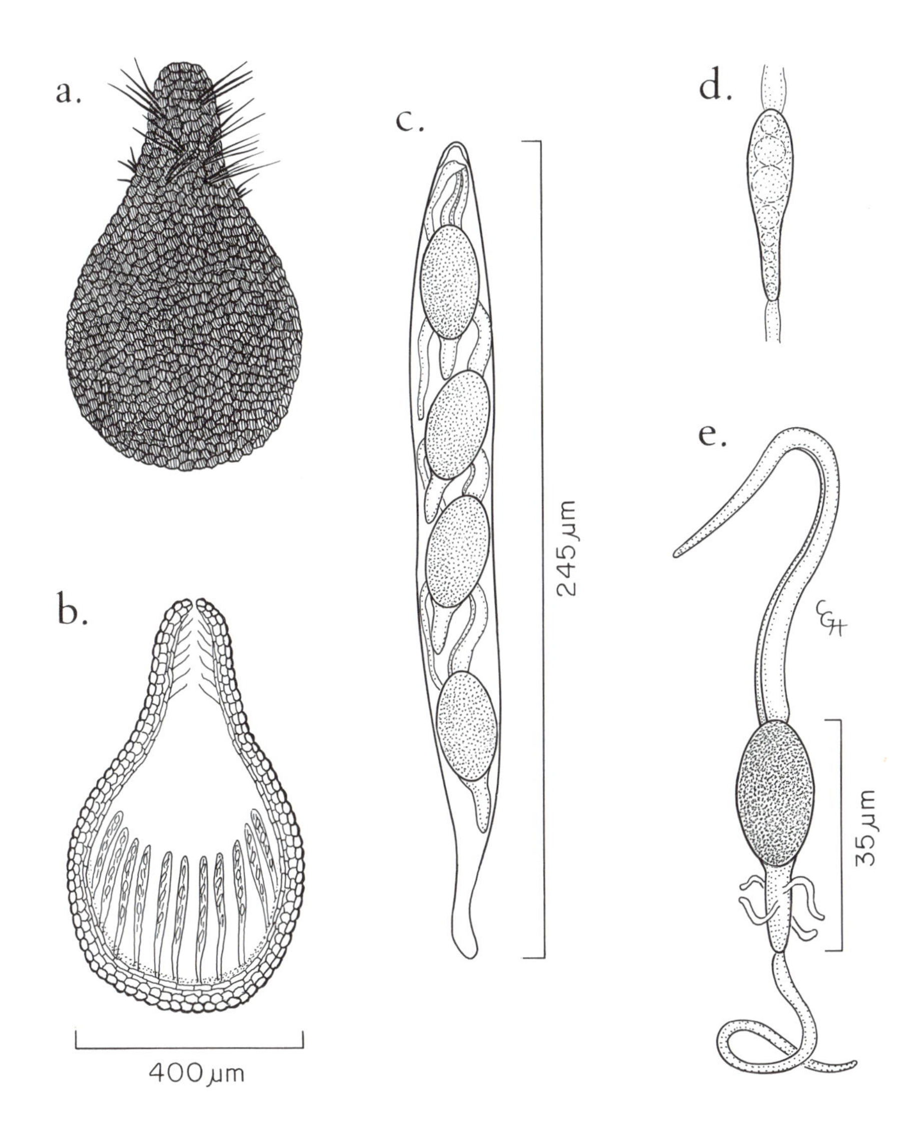

Podospora pauciseta. **a.** Exterior of perithecium. **b.** Section through perithecium with asci. **c.** Ascus with ascospores. **d.** Young one-celled ascospore with gelatinous appendages at each end. **e.** Mature two-celled ascospore with gelatinous appendages.

was a small apiculum that was sometimes lacking; it is also included in *Podospora*. *Podospora pauciseta* (Ces.) Trav. [=*P. anserina* (Ces. ex Rob.) Niessl] has been much used in experimental studies; it is unusual in being secondarily homothallic.

References: Dennis, 1981; Ellis and Everhart, 1892; Furuya and Udagawa, 1972; Lundqvist, 1972; Matsushima, 1971, 1975; Mirza and Cain, 1969; Moreau, 1953; Munk, 1957; BK 334; IM 36:C710.

➔

Fig. 7. **A-C.** *Sordaria fimicola*. **A.** Exterior of perithecium. X109. **B.** Section through perithecium with asci. X121. **C.** Ascus with ascospores. X339. **D-F.** *Venturia liriodendri*. **D.** Section through pseudothecium in leaf. X485. **E.** Ascus with ascospores. X121. **F.** Mature ascospores. X1154. **H-I.** *Apiosporina morbosa*. **H.** Section through ascostroma with ascigerous locules. X89. **I.** Close-up of ascigerous locule with asci. X143.

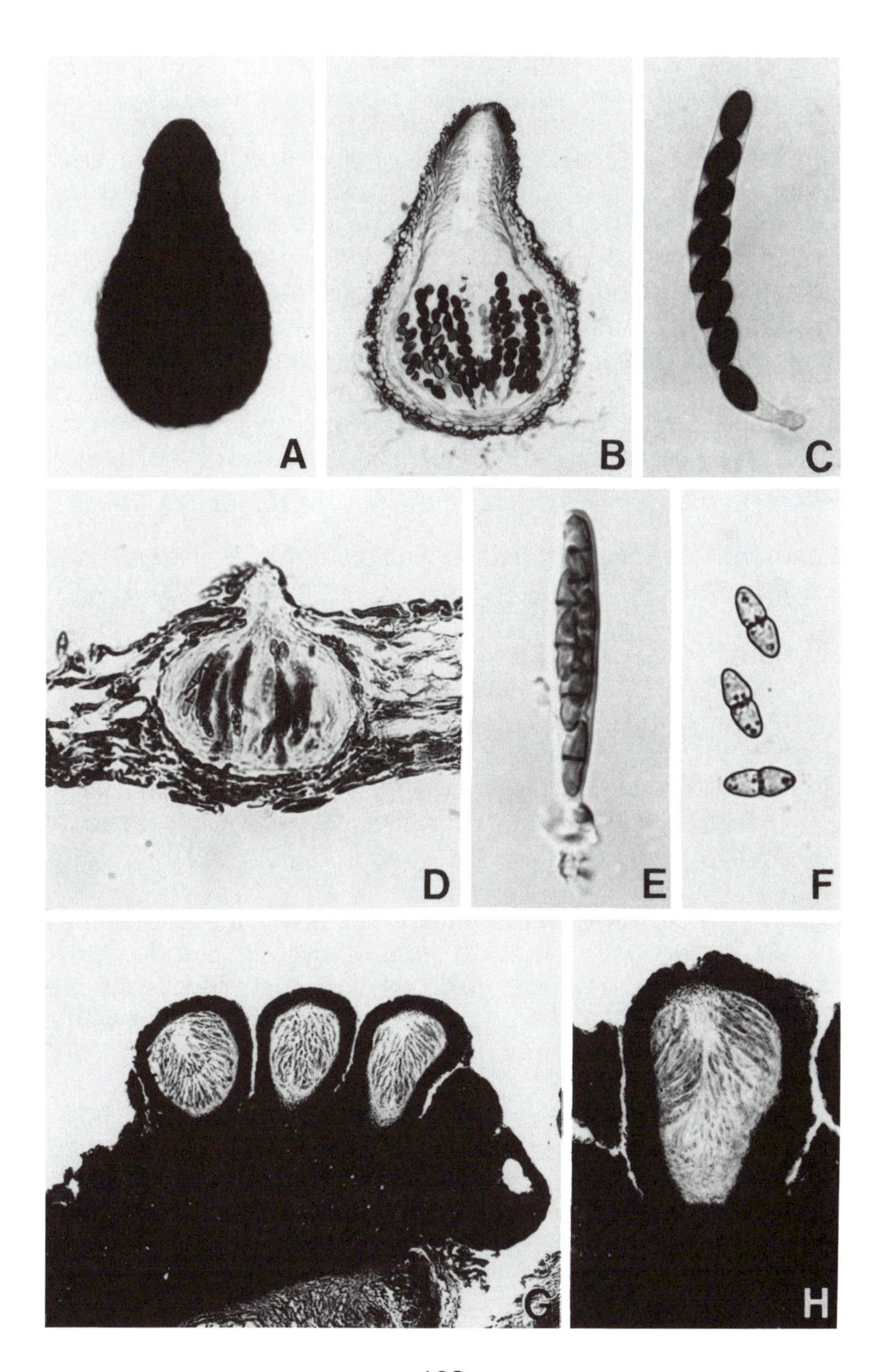
A
B
C
D
E
F
G
H

VENTURIA Sacc.

Ascoma a uniloculate, perithecioid pseudothecium, immersed in host tissue, separate or clustered, with erumpent ostiolar papilla, often with setae around ostiole. Ascomal wall composed of brown, somewhat thick-walled pseudoparenchyma cells, inner wall cells hyaline and thin-walled. Centrum with filamentous, branched, septate, pseudoparaphyses. Asci bitunicate, cylindrical or oblong, short-stalked, 8-spored. Ascospores oblong, elliptical, clavate, or fusoid, olive-green to pale or dark brown, smooth or roughened, sometimes with a thin gelatinous sheath, septate at, above, or below the middle, constricted or not at the septum. (Fig. 7D-F).

Anamorphs: *Cladosporium, Fusicladium, Pollaccia,* and *Spilocaea.*

Habitat: Parasitic on living plant tissues, usually leaves, causing diseases, with ascomata of some species maturing in overwintered leaves.

Representative species: *Venturia inaequalis* (Cooke) G. Wint. (Anam. *Spilocaea pomi* Fr.:Fr.), cause of leaf scab of apple.

Comments: In *Teratosphaeria* Syd. & P. Syd. the ascospores are brown, with a median septum, cylindrical to fusiform, the pseudothecia are glabrous, and the anamorphs are pycnidial. *Didymella* Sacc. has hyaline ascospores with a median septum.

References: Barr, 1968; Dennis, 1981; Ellis and Ellis, 1985; Munk, 1957; Sivanesan, 1977, 1984; FC 36, 181-182, 194, 223-225, 247, 291; CMI 401-402, 404-405, 482-483, 706; IM 2:C23.

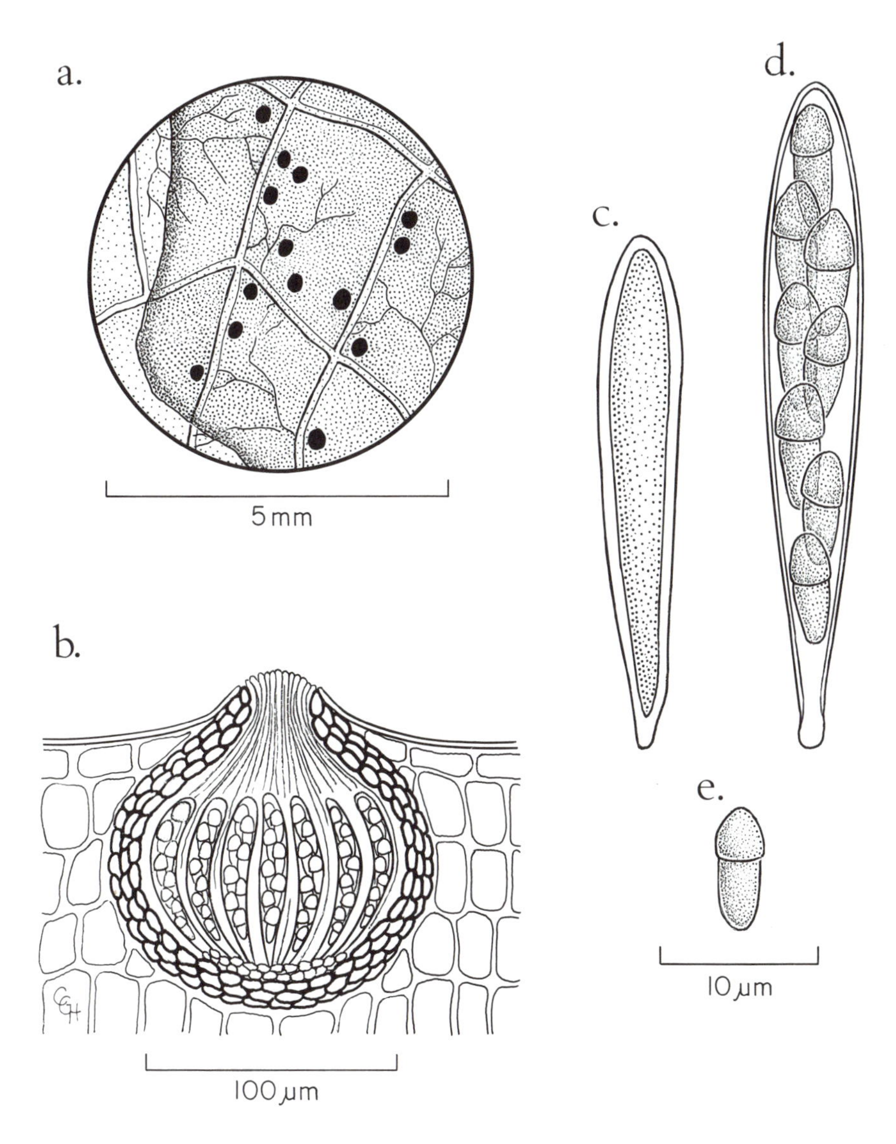

Venturia liriodendri. **a.** Portion of leaf with erumpent ostioles (dark spots). **b.** Section through pseudothecium with asci. **c.** Immature ascus. **d.** Mature ascus with ascospores. **e.** Mature ascospore.

DELITSCHIA Auersw.

Ascoma a uniloculate, perithecioid pseudothecium, scattered or gregarious, partially or entirely immersed in substrate; pseudothecia dark brown to black, smooth or roughened, globose to subglobose or obpyriform; ostiolar neck papillate to long cylindrical. Ascomal wall pseudoparenchymatous, semi-transparent to opaque, membranaceous to coriaceous. Centrum containing pseudoparaphyses. Asci bitunicate, cylindric to clavate, long- or short-stalked, 4-many-spored. Ascospores dark brown to almost black, with an elongated germ slit, oval to broadly fusoid, 2-celled, often constricted at septum, septum median, transverse or oblique, sometimes separating into part-spores, surrounded by a gelatinous sheath.

Anamorph: None reported.

Habitat: All are coprophilous.

Representative species: *Delitschia didyma* Auersw., common on various kinds of dung.

Comments: In *Preussia* Fuckel the ascospores have three or more septa and the pseudothecium is nonostiolate.

References: Cain, 1934; Dennis, 1981; Ellis and Ellis, 1985; Jeng et al., 1977; Luck-Allen and Cain, 1975; Munk, 1957; IM 9:C120.

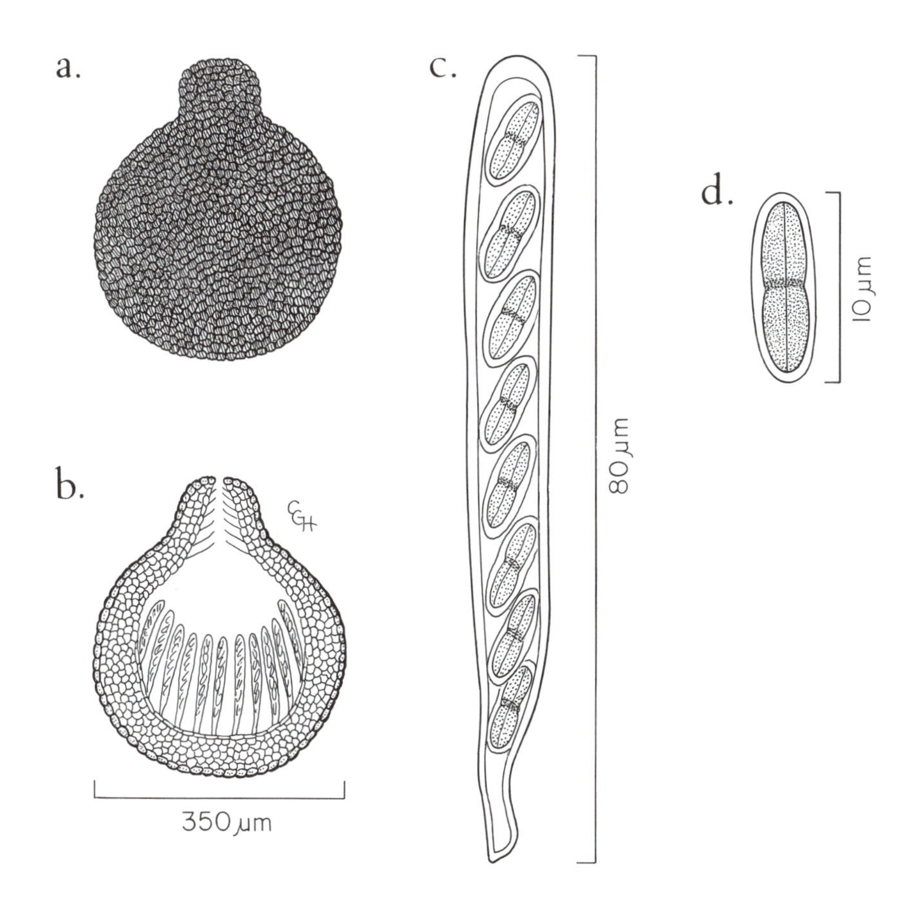

Delitschia marchalii. **a.** Exterior of pseudothecium. **b.** Section through pseudothecium with asci. **c.** Ascus with ascospores. **d.** Mature ascospore with gelatinous sheath.

DIDYMOSPHAERIA Fuckel

Ascoma a uniloculate, perithecioid pseudothecium, scattered or aggregated in groups, immersed and subepidermal, sometimes with slight clypeus development around ostiole. Pseudothecia dark brown, subglobose, with an ostiolar papilla. Ascomal wall composed of small angular cells that are darker around the ostiole. Centrum with branched, filiform, pseudoparaphyses. Asci bitunicate, cylindrical to cylindric-clavate, subclavate or broadly obovoid, short-stalked or subsessile, 1-8-spored. Ascospores 2-celled, septum median, constricted or not, oblique in ascus, smooth or minutely verruculose, olivaceous brown to dark brown.

Anamorphs: *Ascochyta, Dendrophoma*, and *Periconia.*

Habitat: On stems and leaves of herbaceous plants and on seeds.

Representative species: *Didymosphaeria arachidicola* (Chochrjakov) Alcorn, Punith. & McCarthy (Anam. *Ascochyta adzmethica* Schoschiaschvili), cause of web blotch disease of peanut *(Arachis* spp.).

Comments: In *Otthia* Nitschke ex Fuckel the ascoma is erumpent through host tissues, usually on a hypostroma.

References: Dennis, 1981; Ellis and Ellis, 1985; Ellis and Everhart, 1892; Müller and von Arx, 1962; Munk, 1957; Scheinpflug, 1958; Sivanesan, 1984; CMI 736; IM 26:C459.

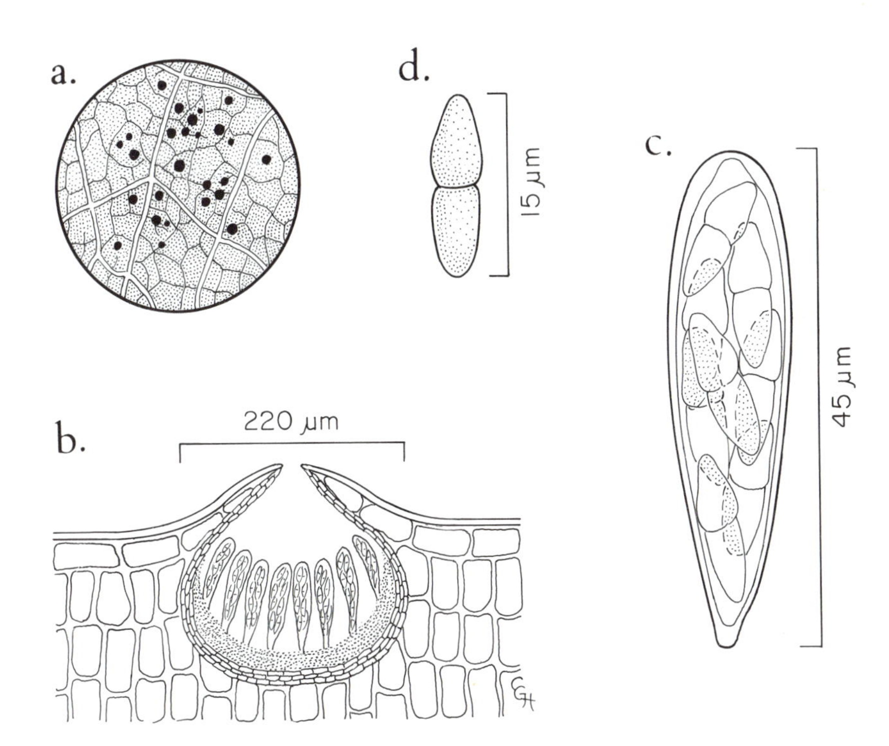

Didymosphaeria arachidicola. **a.** Erumpent pseudothecia on peanut leaf. **b.** Section through pseudothecium. **c.** Ascus with ascospores. **d.** Mature ascospore.

APIOSPORINA Höhn.

Mycelium internal, causing malformation of host tissues, on which a subiculum or pseudoparenchymatous stroma is formed; stroma black, carbonaceous, bearing greenish to olivaceous conidia when young and later developing ascigerous locules on surface. Ascoma black, ostiolate, globose to turbinate, seated on and partly immersed in ascostroma. Ascomal wall pseudoparenchymatous on exterior, lined with flattened, hyaline cells. Centrum containing filamentous pseudoparaphyses. Asci bitunicate, clavate to short-cylindric, 8-spored. Ascospores initially hyaline, becoming greenish to olivaceous or pale brown, obovoid, fusoid or clavate, unequally 2-celled, with septum in lower third, smooth. (Fig. 7G-H).

Anamorphs: *Cladosporium* and *Fusicladium.*

Habitat: Parasitic on woody trees and shrubs.

Representative species: *Apiosporina morbosa* (Schwein.:Fr.) Arx [Anam. *Fusicladium* sp.], cause of black knot disease of cherry and related stone fruits *(Prunus* spp.). This fungus has also been known as *Dibotryon morbosum* (Schw.: Fr.) Theiss. & Syd.

Comments: The genus *Dibotryon* Theiss. & Syd. is considered a synonym of *Apiosporina.* In *Metacoleroa* Petr. the ascospores are greenish to brownish with a median septum.

References: Barr, 1968; Sivanesan, 1984; CMI 224; FC 76, 84.

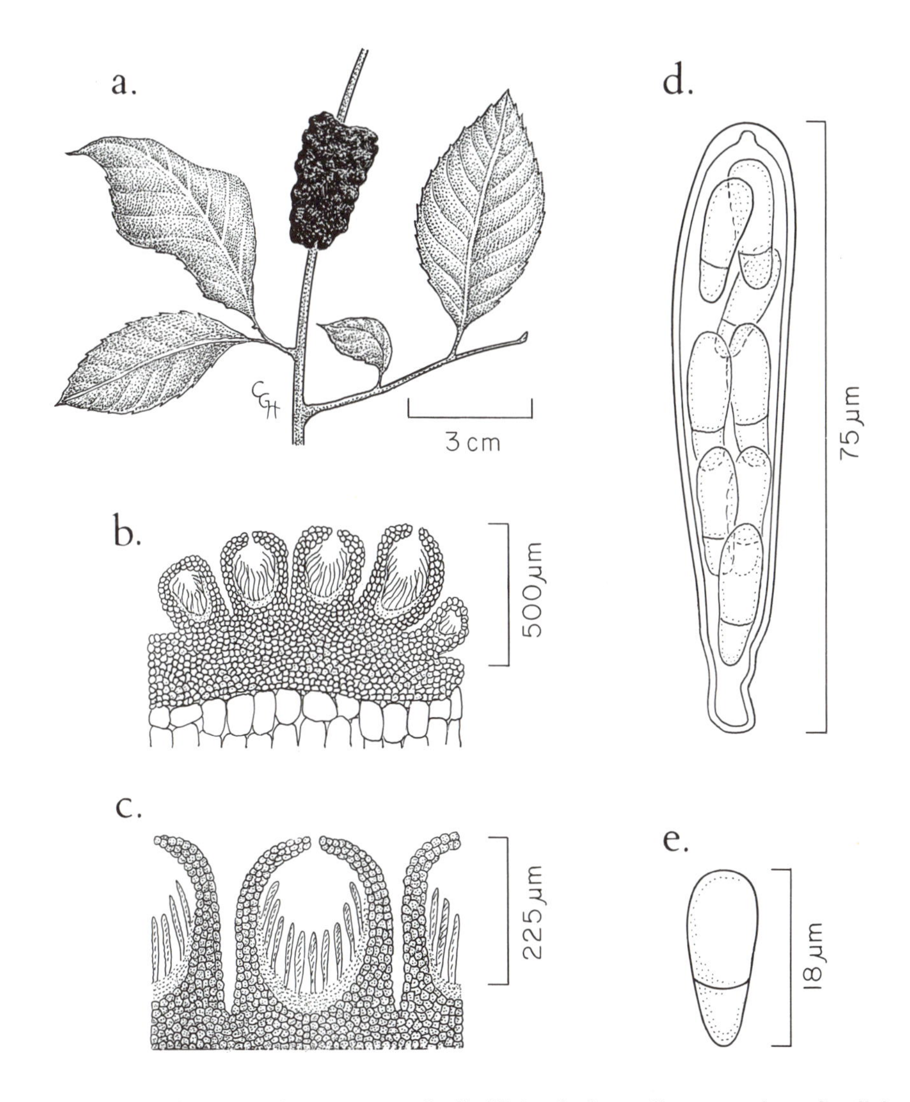

Apiosporina morbosa. **a.** Gall ("black knot") on twig of wild cherry. **b.** Section through gall bearing ascostroma with protuberant ascigerous locules. **c.** Section through ascigerous locule with asci. **d.** Ascus with ascospores. **e.** Mature ascospore.

LEMBOSIA Lév.

Ascoma an elongate, dimidiate pseudothecium, borne on surface of mycelium; mycelium superficial, much-branched, forming dark colonies on host surface; mycelial hyphae dark brown, many-septate, with lateral hyphopodia. Ascostroma linear, sometimes branched and Y-shaped, covered by a black aggregation of hyphae that opens by a longitudinal slit at maturity. Centrum containing branched, filamentous pseudoparaphyses. Asci bitunicate, ellisoid, oblong, clavate or subglobose, thickened at apex, 4-8-spored. Ascospores 2-celled, broadly cylindrical to ovoid, septum median or submedian, becoming brown at maturity. (Fig. 6H-I).

Anamorph: None reported.

Habitat: On living leaves of vascular plants, especially in the tropics.

Representative species: *Lembosia coccolobae* Earle, on *Coccoloba* spp.

Comments: The genus *Morenoella* Speg. is distinguished by the lack of pseudoparaphyses.

References: Arnaud, 1918; Ellis and Ellis, 1985; Müller and von Arx, 1962; IM 8:C111.

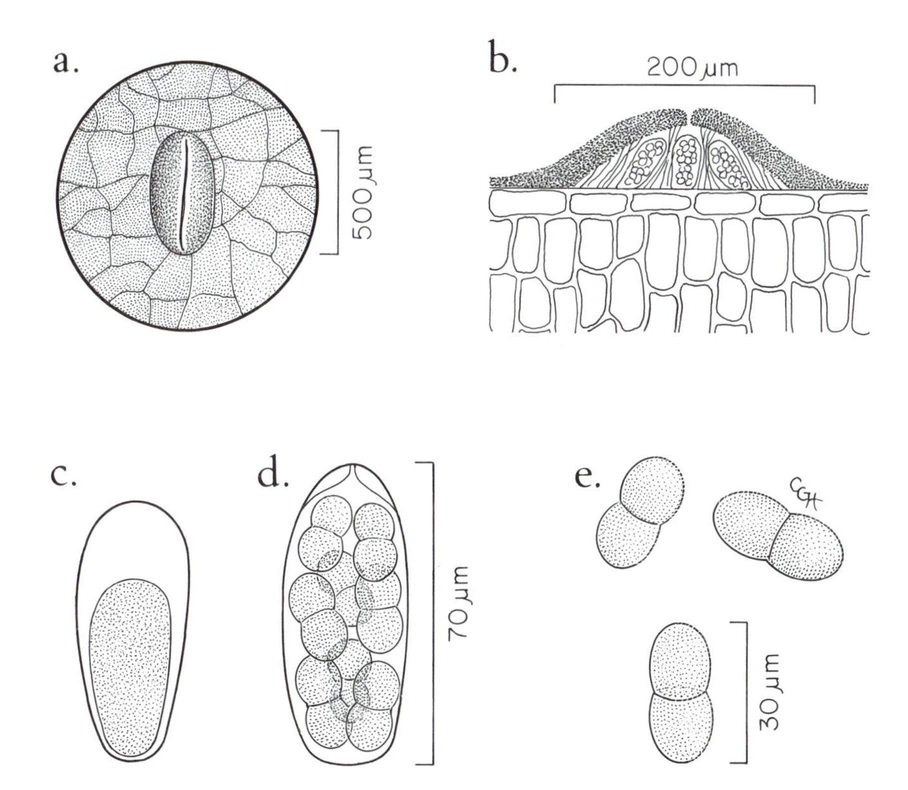

Lembosia melastomum. **a.** Ascoma on leaf of *Miconia.* **b.** Section through ascoma with asci. **c.** Young bitunicate ascus. **d.** Mature ascus with ascospores. **e.** Mature ascospores.

OPHIODOTHELLA (Henn.) Höhn.

Ascoma an ostiolate perithecium, immersed in host tissue, single or clustered, with erumpent ostiolar neck surrounded by a black clypeus; perithecium brown, broadly obpyriform to laterally ovoid, with a single ostiolar neck, but sometimes ostioles amphigenous; ostiole lined with periphyses. Perithecial wall several layers thick, outer cells angular, moderately thick-walled and pigmented, inner cells thin-walled, hyaline and flattened. Centrum containing filamentous, branched paraphyses. Asci unitunicate, lining the perithecial wall, but arranged only at sides of perithecia with amphigenous ostioles, ellipsoidal to oval, short-stalked, with a thickened apex and amyloid pore, 8-spored. Ascospores filiform, 1-celled, lying parallel in ascus or coiled around one another, hyaline, but often appearing greenish, with a single row of oil droplets when freshly discharged.

Anamorph: Unnamed acervular coelomycete.

Habitat: Parasitic in living leaves of flowering plants, causing diseases, ascomata sometimes maturing in overwintered leaves.

Representative species: *Ophiodothella vaccinii* Boyd, cause of leafspot of *Vaccinium arboreum* Marsh.

Comments: In *Gaeumannomyces* Arx & D. Olivier and *Telimenia* Racib. the ascospores are septate; in *Gaeumannomyces* the asci have a refractive ring at the apex, which is lacking in *Telimenia.*

References: Boyd, 1934; Swart, 1982.

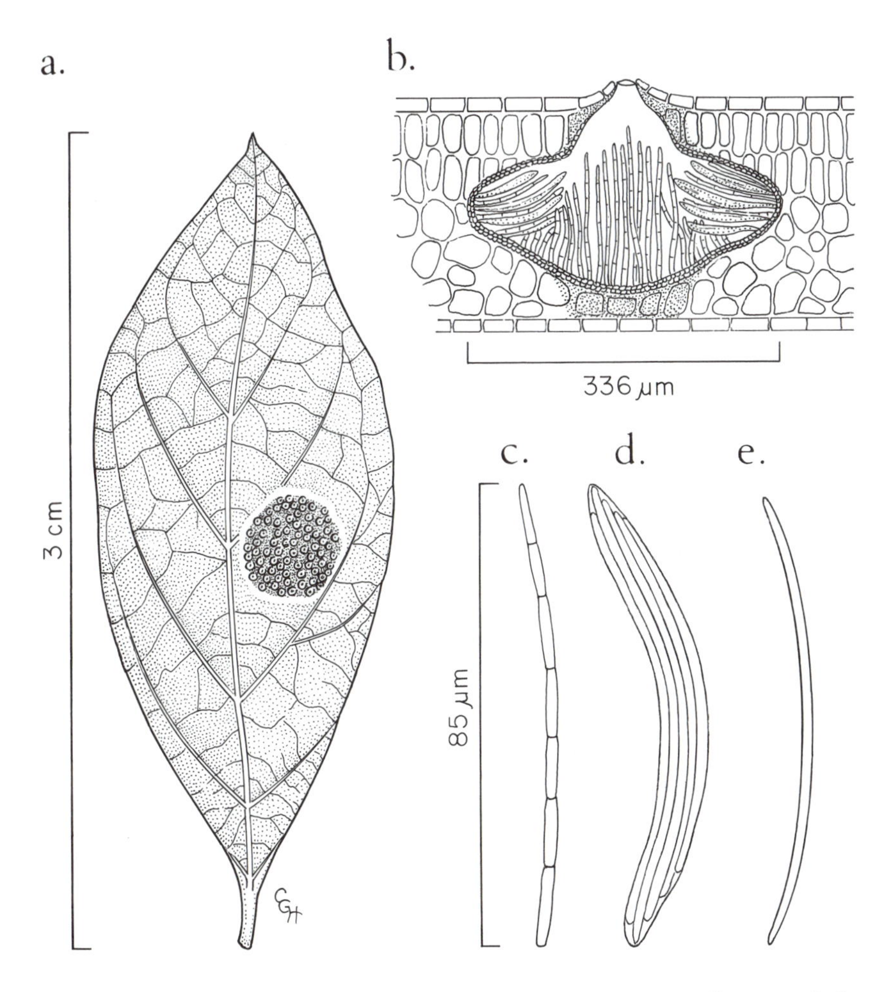

Ophiodothella vaccinii. **a.** Leafspot with protruding ostiolar necks on *Vaccinium* species. **b.** Section through perithecium in host leaf. **c.** Paraphysis. **d.** Ascus with ascospores. **e.** Mature ascospore.

LINOSPORA Fuckel

Ascoma an ostiolate perithecium, formed in host tissues beneath a blackened clypeus and surrounded by a pseudostroma that is delimited by a blackened zone; perithecia single, lying horizontal in leaf, ostiole lateral, curved and extending to leaf surface. Ascomal wall brown. Asci unitunicate, elongate to cylindrical, 4-8-spored. Acospores hyaline, filiform or cylindrical, 1-2-several-celled.

Anamorphs: *Asteroma* and *Melasmia.*

Habitat: In living leaves of deciduous trees, with ascospores maturing after overwintering.

Representative species: *Linospora gleditschiae* J. H. Miller & F. A. Wolf [Anam. *Melasmia gleditschiae* (Lév.) Ellis & Everh.], on living leaves of honey locust (*Gleditsia triacanthos* L.).

Comments: *Ceuthocarpon* P. Karst. has been used for species with a short, broad ostiolar neck, but it is now considered a synonym of *Linospora*. *Hypospilina* (Sacc.) Traverso differs in having apiosporous, ellipsoid to ovoid ascospores.

References: Barr, 1978; Kojwang and Kurkela, 1984; Miller and Wolf, 1936; Munk, 1957; IM 5:C71.

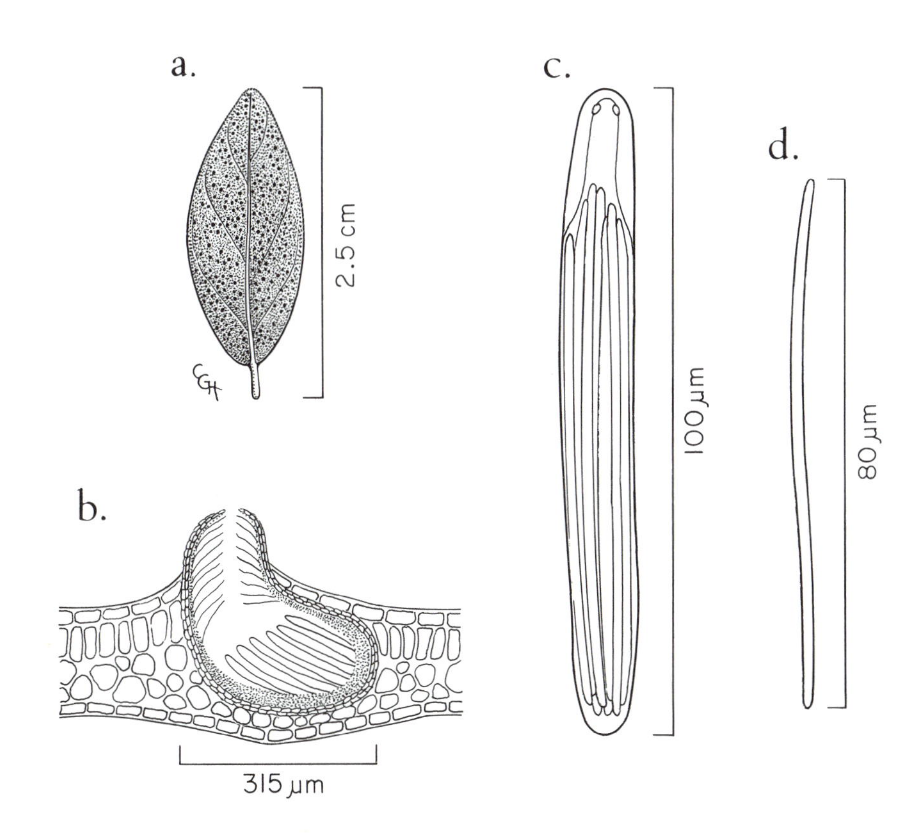

Linospora gleditschiae. **a.** Erumpent ostioles on leaflet of honey locust. **b.** Section through perithecium with asci. **c.** Ascus with ascospores. **d.** Mature ascospore.

GAEUMANNOMYCES **Arx & D. Olivier**

Mycelium superficial, brown, forming a thin mat on surface of host tissues; hyphae with simple or lobed, dark brown hyphopodia. Ascoma an ostiolate perithecium, solitary, immersed, black, globose to broadly oval; ostiolar neck short, broad and oblique. Ascomal wall composed of flattened cells, outer cells dark brown and thick-walled, inner cells hyaline and thinner-walled. Centrum with filamentous paraphyses. Asci unitunicate, with a refractive, nonamyloid, apical ring, elongate-clavate, 8-spored. Ascospores filiform to narrowly clavate, tapering toward the ends, straight or curved, several-septate, hyaline or yellowish.

Anamorph: *Phialophora.*

Habitat: On roots, stems and leaves of grasses and sedges.

Representative species: *Gaeumannomyces graminis* (Sacc.) Arx & D. Olivier, cause of take-all disease of wheat (*Triticum* spp.) and other cereals. This species has long been known as *Ophiobolus graminis* (Sacc.) Sacc.

Comments: *Linocarpon* Syd. & P. Syd. differs in having a small clypeus around the ostiolar necks and in the consistently upright perithecia with central necks. In *Ophiobolus* Riess the asci are bitunicate and there is a swollen cell on either side of the central septum of the ascospore.

References: Domsch et al., 1980; Ellis and Ellis, 1985; Walker, 1972, 1980; CMI 381- 383; FC 37; IM 5:C62.

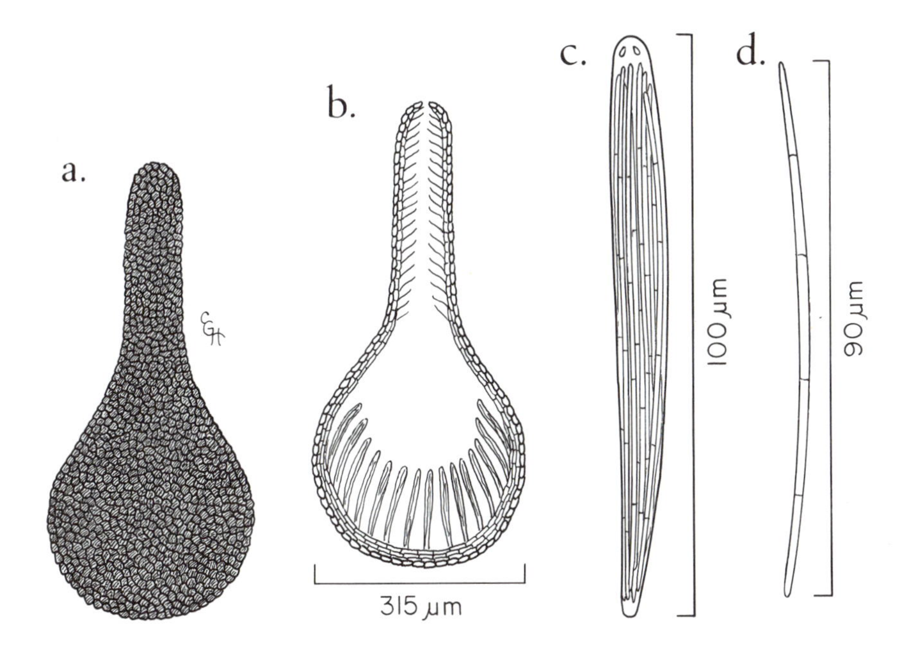

Gaeumannomyces graminis. **a.** Exterior of perithecium. **b.** Section through perithecium with asci. **c.** Ascus with ascospores. **d.** Mature ascospores.

COCHLIOBOLUS Drechs.

Ascoma a uniloculate, perithecioid pseudothecium, formed free or on a column or flat stroma; pseudothecium dark brown to black, globose, ellipsoid or obpyriform, ostiolate, with long or short neck; neck often bearing brown hyphae. Ascomal wall composed of pseudoparenchyma cells. Centrum containing filiform, septate, branched pseudoparaphyses. Asci bitunicate, cylindrical to broadly clavate, 1-8-spored. Ascospores filiform and septate, hyaline, but sometimes becoming light brown, and sometimes with a gelatinous sheath, helically coiled in the ascus.

Anamorphs: *Curvularia* and *Drechslera.*

Habitat: On herbaceous plants, often causing diseases.

Representative species: *Cochliobolus carbonum* Nelson [Anam. *Bipolaris zeicola* (G. L. Stout) Shoemaker], cause of leafspot disease of maize (*Zea mays* L.).

Comments: The genus *Pseudocochliobolus* Tsuda, Ueyama & Nishihara has been used for species with loosely coiled ascospores in pseudothecia that form on a basal column or stroma, but Alcorn considers it a synonym of *Cochliobolus.*

References: Alcorn, 1982, 1983; Domsch et al., 1980; Sivanesan, 1984; CMI 301-302, 341, 349, 473-474, 701-703, 726, 728; IM 4:C41.

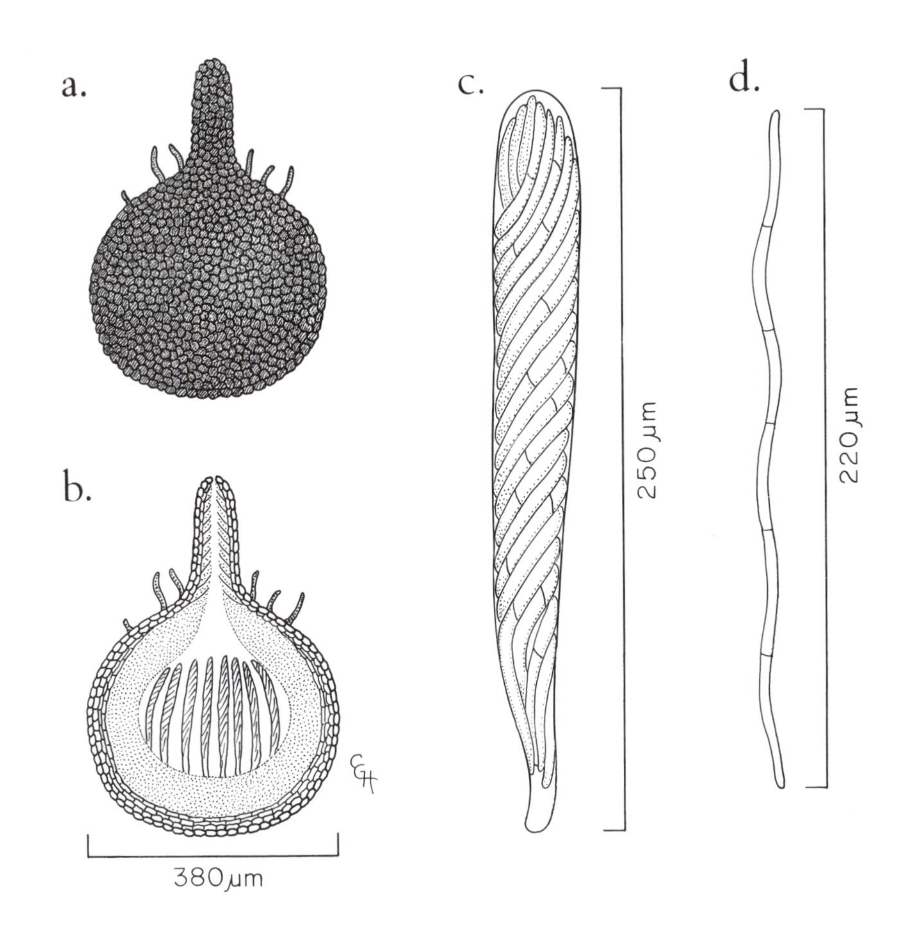

Cochliobolus carbonum. **a.** Exterior of pseudothecium. **b.** Section through pseudothecium with asci. **c.** Ascus with spirally arranged ascospores. **d.** Mature ascospore.

BALANSIA **Speg.**

Ascomata stromatic, with stroma originating from systemic mycelium in host tissues, first forming a thin, flat hypothallus on leaf surface, hyaline at first, becoming reddish-brown to light, then dark brown. Conidia are produced on the hypothallus. Perithecial stroma forming later, over hypothallus, effuse to elongate-pulvinate, or hemispherical and sessile, or stipitate-capitate, dark brown to blackish on surface, interior hyaline, fleshy. Perithecia immersed in a single layer in periphery of stroma, crowded, ovate, ostiolate, with ostiolar necks protruding slightly. Ascomal wall thin, composed of flattened cells. Centrum containing lateral paraphyses. Asci unitunicate, cylindrical with short stalk, with thickened apical cap, formed in a basal cluster, 8-spored. Ascospores filiform, hyaline, several-septate, often breaking into part-spores. (Fig. H-J).

Anamorph: *Ephelis.*

Habitat: Parasitic on grasses, causing various diseases.

Representative species: *Balansia epichloë* (Weese) Diehl, an effuse species, and *B. claviceps* Speg., a tropical stipitate-capitate species.

Comments: The species of *Balansia* with effuse stromata have been placed in *Dothichloë* Atk., which is considered a synonym of *Balansia*. *Atkinsoniella* Diehl differs from *Balansia* in having both typhodial and ephelidial anamorphs.

References: Diehl, 1950; Viégas, 1944; CMI 640.

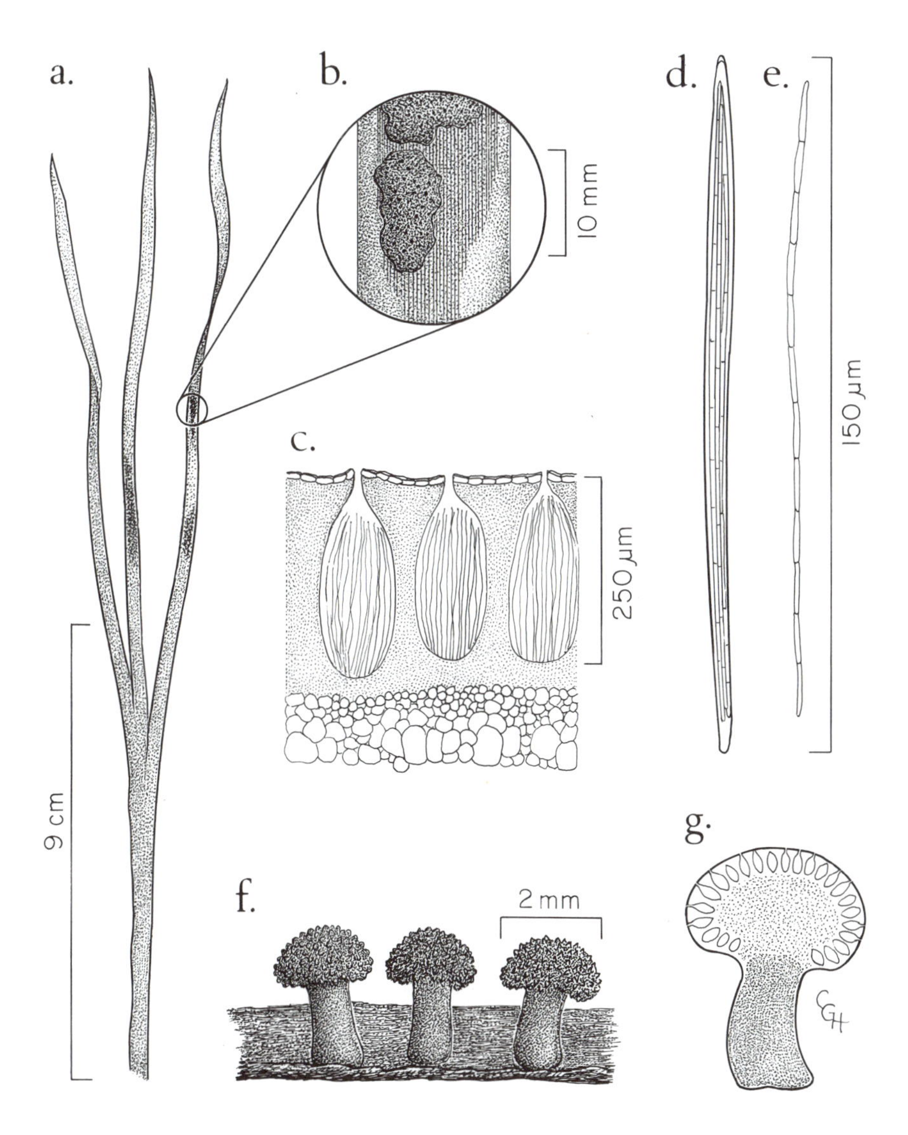

Balansia henningsiana. **a.** Stromata on grass leaves. **b.** Close-up of perithecial stroma. **c.** Section through stroma with perithecia. **d.** Ascus with ascospores. **e.** Mature ascospore. *Balansia claviceps.* **f.** Perithecial stromata on grass leaf. **g.** Section through perithecial stroma.

EPICHLOE (Fr.) Tul. & C. Tul.

Ascoma formed in a stroma composed of fungal tissues; stroma effuse, superficial on host, often encircling stem, initially white and byssoid, becoming fleshy and brightly colored. Ascoma an ostiolate perithecium, immersed in stroma, but often with ostiolar necks free, obpyriform. Ascomal wall thin, composed of elongated, thin-walled cells. Centrum containing lateral paraphyses. Asci unitunicate, long cylindrical, with a thickened, nonamyloid apex, 8-spored. Ascospores hyaline, filiform, multiseptate, not fragmenting.

Anamorph: *Acremonium* (*Sphacelia*).

Habitat: Parasitic on grasses.

Representative species: *Epichloë typhina* (Pers.:Fr.) Tul.

Comments: In *Balansia* Speg. the stromata are brown or black and the anamorph is ephelidial; *Atkinsoniella* Diehl has both ephelidial and typhodial anamorphs.

References: Dennis, 1981; Doguet, 1960; Ellis and Ellis, 1985; Ellis and Everhart, 1892; Munk, 1957; BK 315; CMI 639.

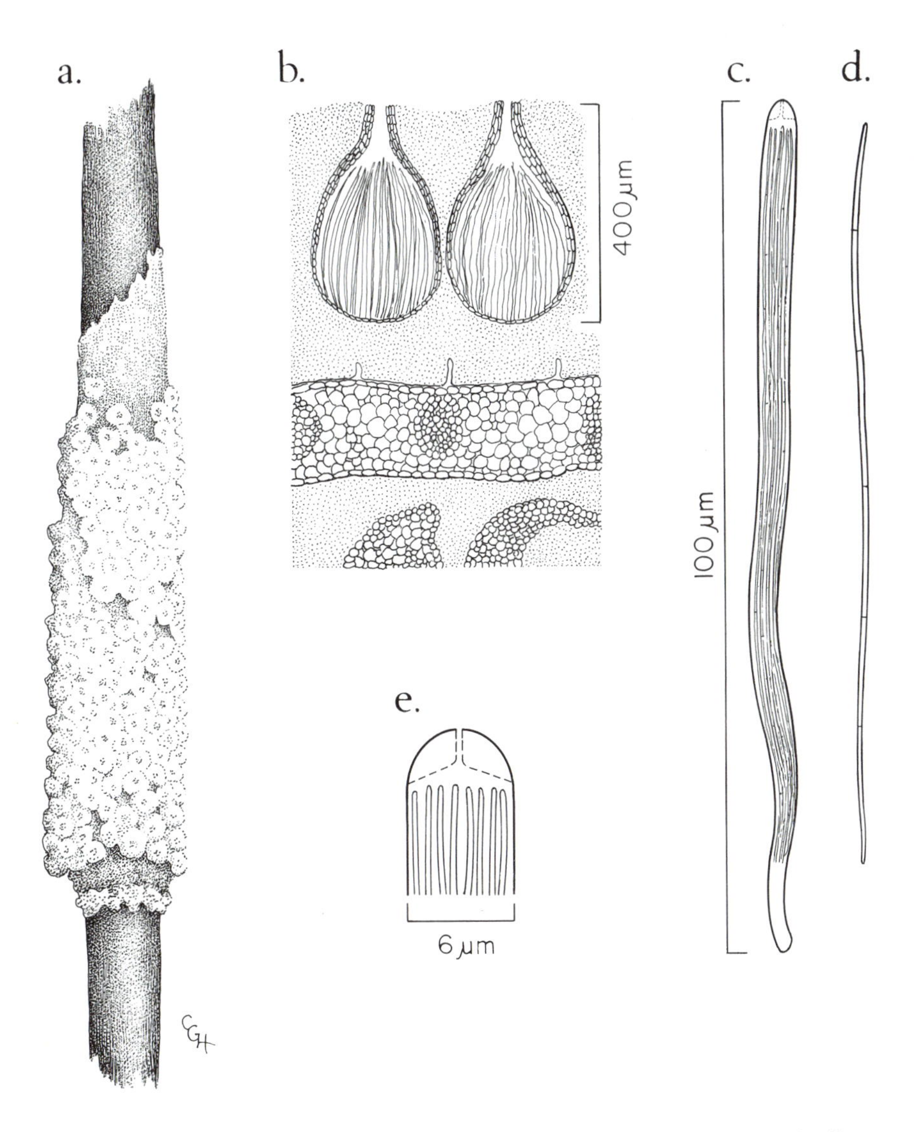

Epichloë typhina **a.** Perithecia seated on stroma encircling grass stem. **b.** Section through stroma with perithecia and asci. **c.** Ascus with ascospores. **d.** Mature ascospore. **e.** Apex of ascus.

MYRIOGENOSPORA Atk.

Mycelium superficial on host tissues, forming two narrow, linear black stromata parallel to long axis of leaf. Ascomata formed in a single row in stroma; ascoma an ostiolate perithecium, immersed in stroma, with slightly protruding ostiolar papilla. Ascomal wall thin, composed of elongated prosenchyma cells. Centrum containing lateral paraphyses with filamentous apices and vesiculose basal cells. Asci unitunicate, formed in a basal cluster, elongated fusoid, with a dome-shaped apical cap, 8-spored. Ascospores hyaline, filamentous and septate, soon separating into part-spores that become fusoid. (Fig. 4F-H).

Anamorph: *Ephelis.*

Habitat: Parasitic on grasses.

Representative species: *Myriogenospora atramentosa* (Berk. & M. A. Curtis) Diehl, cause of tangle-top disease of grasses.

Comments: In *Balansia* Speg. and *Epichloë* (Fr.) Tul. & C. Tul. the ascus is cylindrical; in *Balansia* the stroma is brown or black and in *Epichloë* it is brightly colored.

References: Luttrell and Bacon, 1977; Viégas, 1944.

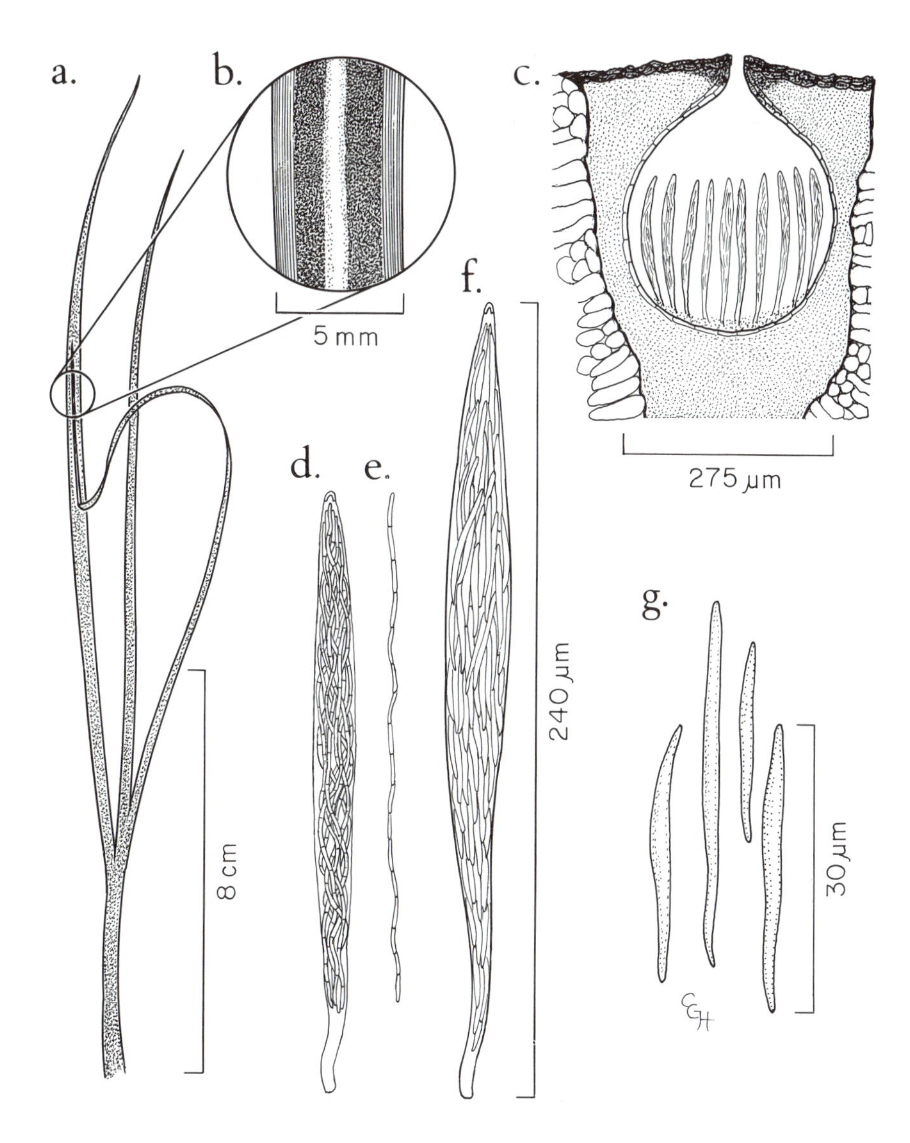

Myriogenospora atramentosa. **a.** Grass stem showing symptoms of "tangle-top" disease. **b.** Close-up of grass leaf with linear stromata. **c.** Section through stroma with perithecium. **d.** Young ascus with ascospores. **e.** Ascospore. **f.** Older ascus filled with part-spores. **g.** Part-spores.

CLAVICEPS Tul.

Ascoma formed in a stroma; stroma arising from a hard, black sclerotium composed entirely of fungal tissues, erect, capitate-stipitate, fleshy, bright- or light-colored, ascomata confined to capitate portion. Ascoma an ostiolate perithecium, immersed in stroma with ostiolar necks protruding slightly, long-obpyriform; ascomal wall thin, composed of thin-walled, elongated cells. Centrum containing lateral paraphyses. Asci unitunicate, long, cylindrical, with a nonamyloid, thickened apical cap, 8-spored. Ascospores hyaline, filiform, multiseptate upon discharge.

Anamorph: *Sphacelia.*

Habitat: Parasitic on grasses and sedges.

Representative species: *Claviceps purpurea* (Fr.:Fr.) Tul., cause of ergot disease of grains.

Comments: *Cordyceps* (Fr.) Link differs in that the perithecial stromata arise from host tissues and no sclerotia are formed; the two genera also have very different hosts.

References: Dennis, 1981; Ellis and Ellis, 1985; Ellis and Everhart, 1892; Langdon, 1954; BK 308; IM 4:C34.

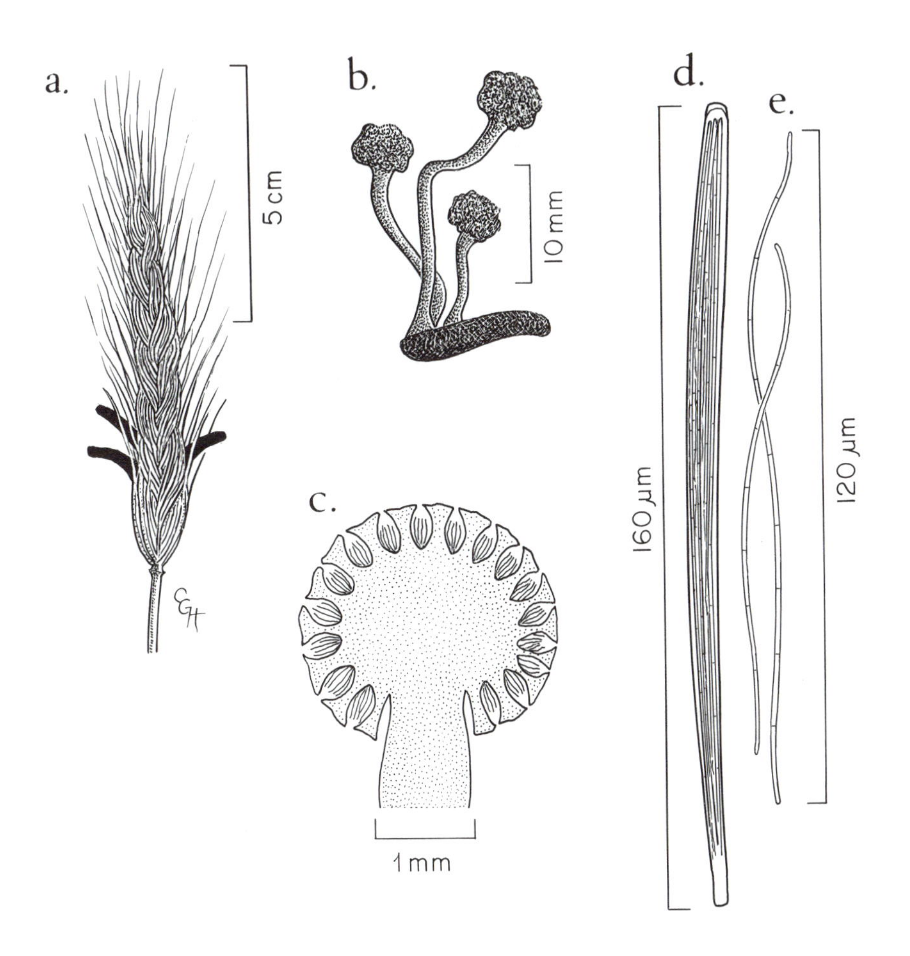

Claviceps purpurea. **a.** Sclerotia on head of rye. **b.** Perithecial stromata arising from sclerotium. **c.** Section through head of stroma with perithecia. **d.** Ascus with ascospores. **e.** Mature ascospores.

CORDYCEPS (Fr.) Link

Ascoma formed in or on a stroma formed from mycelial masses in insects, spiders or ascomata of *Elaphomyces*, cylindric, clavate or capitate-stipitate, simple or sometimes branched, variously colored, white, yellow, orange, red to brown or black. Ascoma an ostiolate perithecium, superficial and free or partly or completely immersed in a fleshly stroma composed entirely of fungal tissue, obpyriform, ovoid, conoid, or subglobose. Centrum containing lateral paraphyses. Asci unitunicate, cylindrical, subfusoid, or narrowly clavate, usually with a thickened, nonamyloid, apical cap, 8-spored. Ascospores filiform, hyaline, multiseptate, often separating into 1-celled part-spores.

Anamorphs: *Akanthomyces, Cephalosporium, Hirsutella, Hymenostilbe, Isaria, Spicaria, Sporotrichum, and Stilbella*.

Habitat: Mostly entomogenous, but some fungicolous.

Representative species: *Cordyceps capitata* on ascomata of *Elaphomyces* spp. and *C. militaris* (Fr.) Link on pupae of Lepidoptera.

Comments: *Ophiocordyceps* Petch has been used for species in which the ascospores do not fragment into part-spores, and *Racemella* Ces. has been used for species with superficial perithecia. Both are regarded as synonyms of *Cordyceps.*

References: Dennis, 1981; Ellis and Everhart, 1892; Kobayasi and Shimizu, 1983; Mains, 1957, 1958; Munk, 1957; Viégas, 1944; BK 309-313.

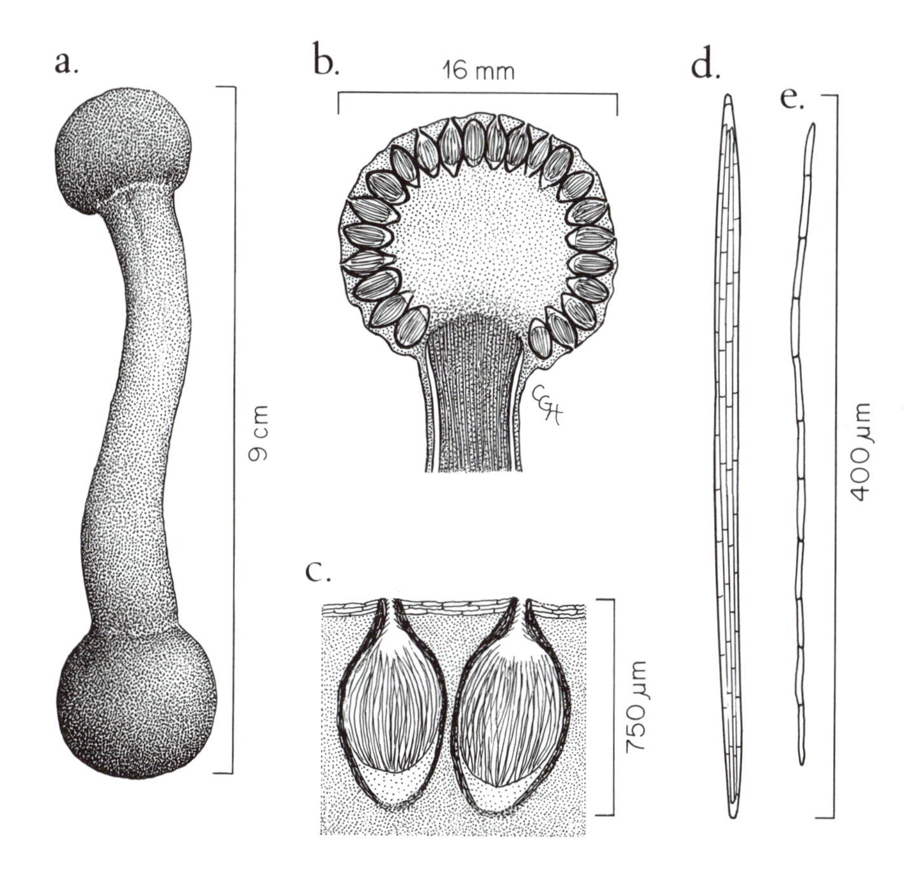

Cordyceps capitata. **a.** Perithecial stroma arising from ascoma of *Elaphomyces.* **b.** Section through head of stroma with perithecia. **c.** Close-up of perithecia with asci. **d.** Ascus with ascospores. **e.** Mature ascospore.

LOPHODERMIUM Chev.

Ascoma elliptical, apothecioid, immersed in host tissues, becoming erumpent at maturity, opening by a longitudinal slit; hymenium flat to slightly concave. Paraphyses simple, filiform. Asci unitunicate, narrowly cylindric to cylindric-clavate, 4-8-spored. Ascospores filiform, 1-celled, hyaline, surrounded by a gelatinous sheath.

Anamorphs: *Labrella* and *Leptostroma.*

Habitat: On conifer needles and occasionally leaves of vascular plants.

Representative species: *Lophodermium pinastri* (Schrad.:Fr.) Chev., widespread on needles of pine (*Pinus* spp.).

Comments: Species with septate, filiform ascospores are placed in *Lophomerum* Ouellette & Magasi.

References: Cannon and Minter, 1986; Darker, 1932, 1967; Dennis, 1981; Ellis and Ellis, 1985; Hunt and Ziller, 1978; Minter, 1981; Minter and Sharma, 1982; Tehon, 1935; BK 294-295; CMI 197, 563-568, 784-789, 795- 798; FC 195; IM 4:C48.

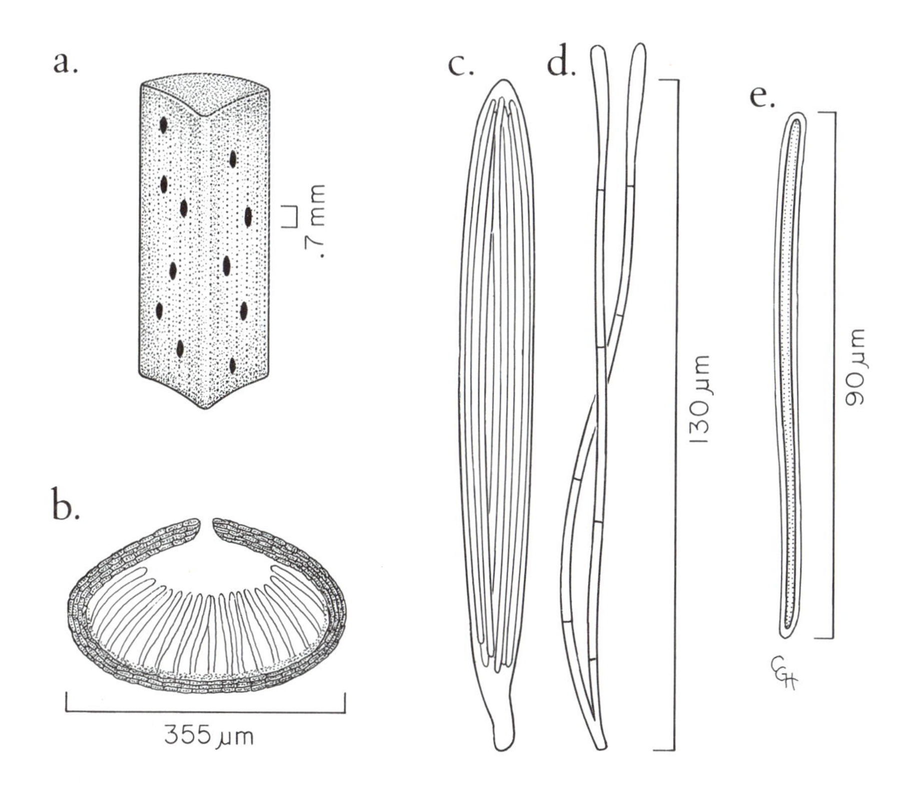

Lophodermium pinastri. **a.** Erumpent apothecia on pine needle. **b.** Section through apothecium with asci. **c.** Ascus with ascospores. **d.** Paraphyses. **e.** Mature ascospore with gelatinous sheath.

RHYTISMA Fr.:Fr.

Ascoma an apothecium formed in a multilocular stroma embedded in host tissues; stroma circular to oval, up to 20 mm diameter, erumpent through host epidermis, exterior black and shiny, interior white, containing numerous hymenial areas, opening by a slit above each hymenium. Paraphyses slender, curled and sometimes forked at tip. Asci unitunicate, clavate, with a conical apex, 8-spored. Ascospores filiform, but sometimes slightly broader at upper end, hyaline, one-celled.

Anamorph: *Melasmia.*

Habitat: Parasitic in living leaves of deciduous trees.

Representative species: *Rhytisma acerinum* (Pers.:Fr.) Fr. (Anam. *Melasmia acerina* Lév.), cause of tar spot disease of maple (*Acer* spp.).

Comments: In *Coccomyces* De Not. the ascoma is covered by distinctly blackened fungal tissue which usually opens in a stellate fashion and the ascospores may be septate or not; in *Propolis* (Fr.) Corda the ascoma is covered mainly by host tissue; and in *Colpoma* Wallr. the ascoma opens by a longitudinal slit. All of these genera form solitary ascomata in host tissues.

References: Cannon and Minter, 1986; Dennis, 1981; Ellis and Ellis, 1985; CMI 791; BK 291; IM 21:C307.

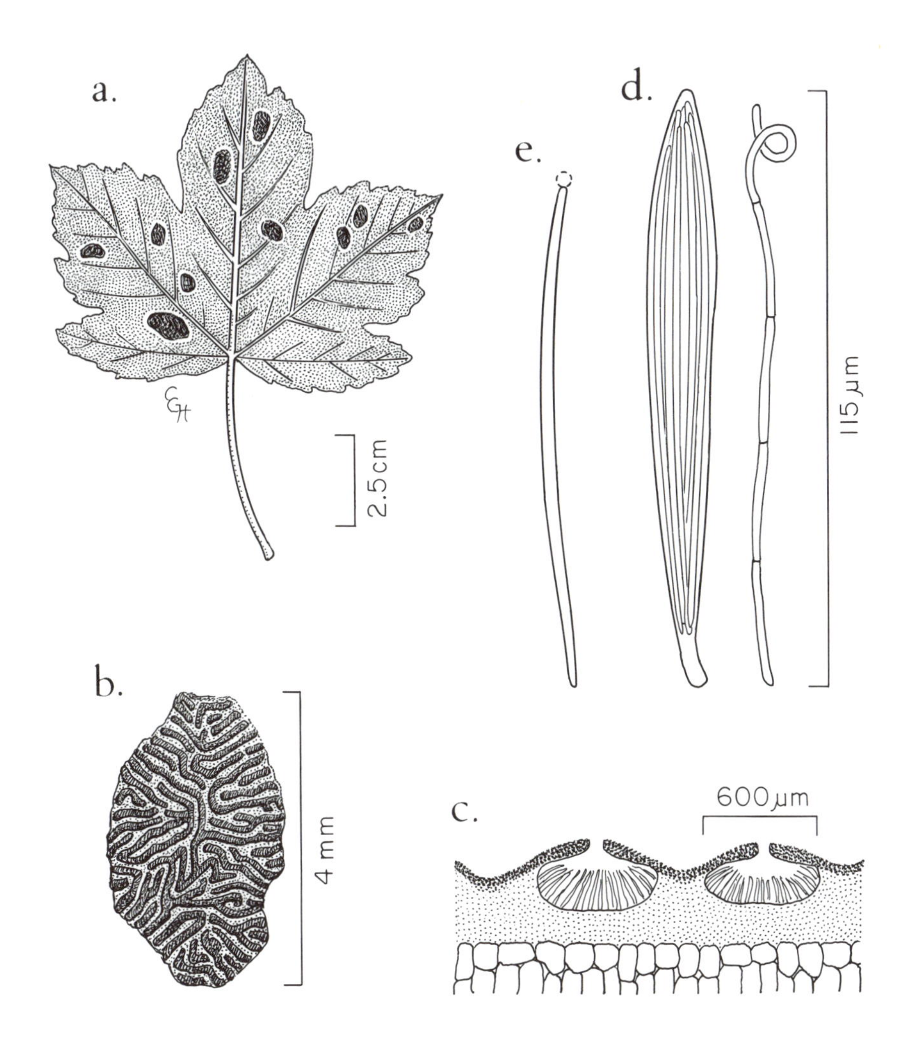

Rhytisma acerina. **a.** Erumpent apothecial stroma on maple leaf. **b.** Close-up of apothecial stroma with fissures in surface. **c.** Section through stroma with two apothecia. **d.** Paraphysis and ascus with ascospores. **e.** Mature ascospore with gelatinous attachment at tip.

LOPHIUM Fr.

Ascoma an erect, mussel-shaped pseudothecium with a distinct basal keel, single, superficial, black, opening by a narrow longitudinal slit. Centrum containing pseudoparaphyses. Asci bitunicate, cylindrical, 8-spored. Ascospores filiform, septate, lying parallel in ascus or spirally coiled, hyaline, but sometimes brownish in age.

Anamorph: None reported.

Habitat: On stems and branches of conifers.

Representative species: *Lophium mytilinum* (Pers.) Fr.

Comments: In *Glyphium* Nitschke ex Lehm. the ascospores are filiform and brown, but the ascoma is hatchet-shaped; in *Mytilinidion* Sacc. the ascospores are phragmosporous, long-fusoid, and brown.

References: Dennis, 1981; Ellis and Ellis, 1985; Zogg, 1962; IM 6:C74.

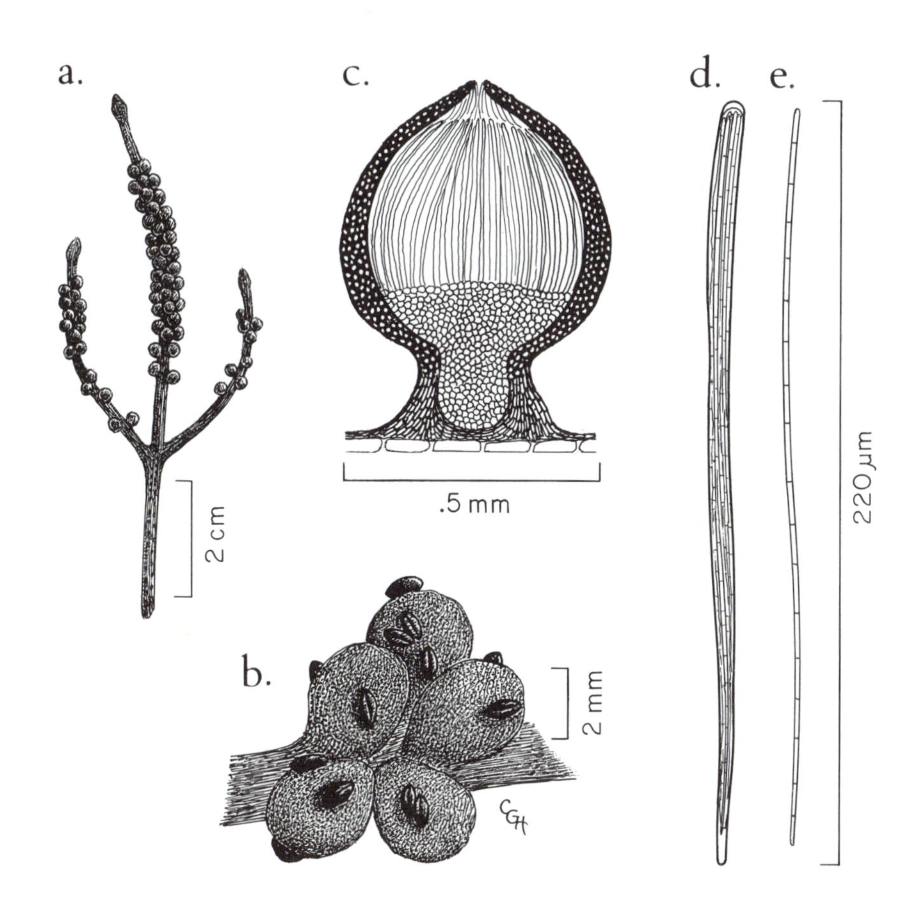

Lophium mytilinum. **a.** Male cones of balsam fir. **b.** Ascomata on cones. **c.** Section through ascoma. **d.** Ascus with ascospores. **e.** Mature ascospore.

SPATHULARIA **Pers.**

Ascoma erect, with a flattened, spathulate or fan-shaped hymenium-bearing receptacle borne on a stipe, up to 10 cm high; receptacle fleshy to leathery, up to 3 cm wide, yellowish to brownish-yellow; stipe terete or somewhat compressed, brown, darker than receptacle, composed of thin-walled hyphae. Paraphyses filiform, branched, curved or coiled at apices. Asci unitunicate, clavate, apex nonamyloid, 8-spored. Ascospores filiform, hyaline, multiseptate, parallel in ascus, sometimes twisted to curved.

Anamorph: None reported.

Habitat: On soil or rotting logs.

Representative species: *Spathularia flavida* Fr.

Comments: In *Spathulariopsis* Maas G. the outer tissues of the stipe are pseudoparenchymatous and the medullary hyphae are thick-walled.

References: Dennis, 1981; Maas Geesteranus, 1972; Mains, 1955; Seaver, 1951; BK 141; IM 10:C146.

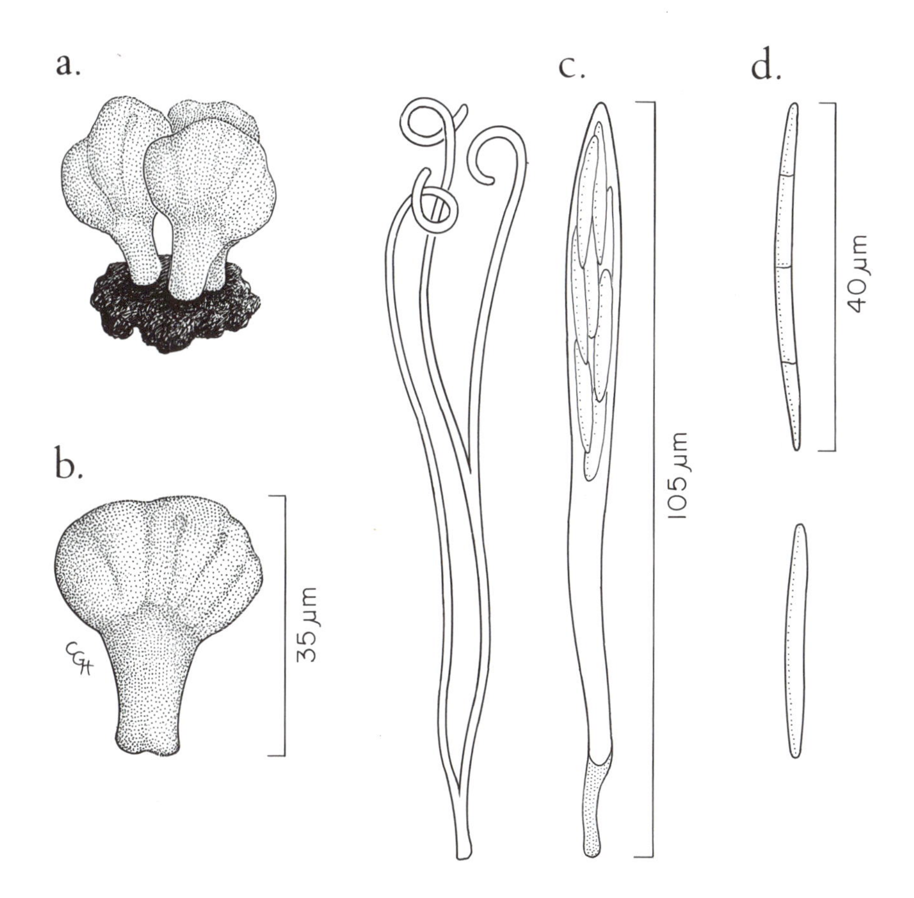

Spathularia flavida. **a.** Cluster of ascomata. **b.** Single ascoma. **c.** Paraphyses and ascus with asccospores. **d.** Top, mature ascospore; bottom, young ascospore.

CALONECTRIA **De Not.**

Ascoma an ostiolate perithecium, solitary to aggregated, superficial, brightly colored, turning purple in KOH; ascomal wall composed of two regions, the outer region with angular to globose, thick-walled, pigmented cells, the cells of the inner region elongate, hyaline and thin-walled. Centrum containing apical paraphyses. Asci unitunicate, broadly clavate to obovate, undifferentiated at apex. Ascospores elliptic to fusiform, often somewhat curved, 1-several-septate, hyaline.

Anamorph: *Cylindrocladium.*

Habitat: On vascular plants, often parasitic.

Representative species: *Calonectria pyrochroa* (Desmaz.) Sacc. [=*C. crotalariae* (Loos) D. K. Bell & Sobers] [Anam. *Cylindrocladium ilicicola* (Hawley) Boedijn & Reitsma (=*C. crotalariae* (Loos) D. K. Bell & Sobers)], pathogenic on numerous vascular plants.

Comments: *Calonectria* has traditionally been used for brightly colored perithecial fungi with hyaline, phragmosporous ascospores, but Rossman restricts the genus to species with a characteristic wall structure and a *Cylindrocladium* anamorph.

References: Bell and Sobers, 1966; Ellis and Ellis, 1985; Ellis and Everhart, 1892; Matsushima, 1971; Rossman, 1979a,b; CMI 421, 423-424, 426, 429-430.

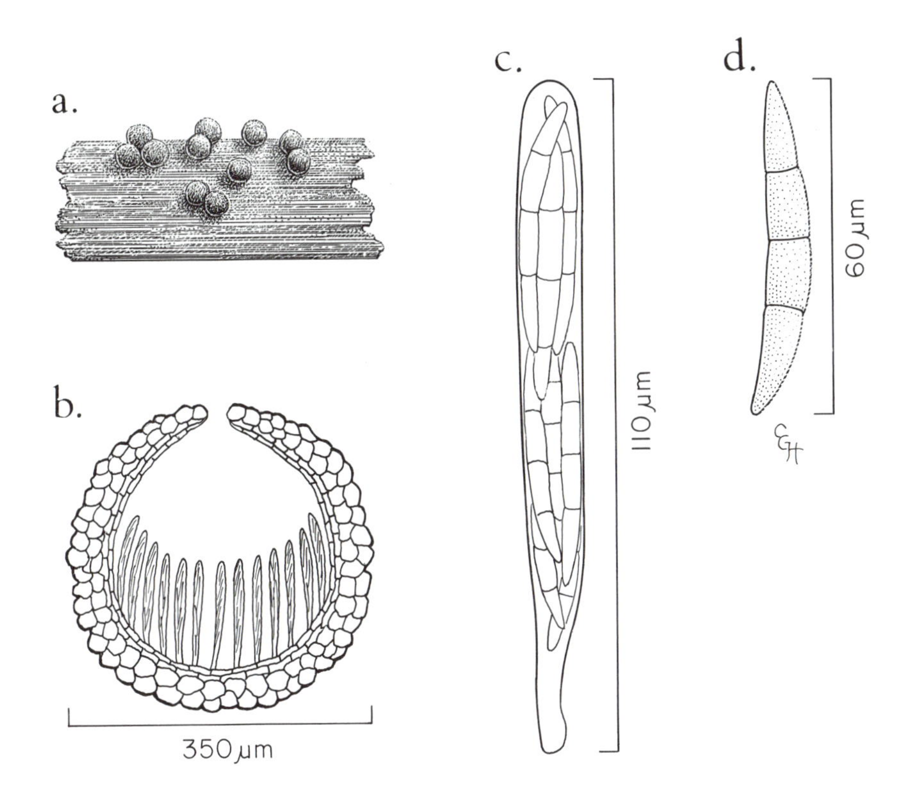

Calonectria pyrochroa. **a.** Perithecia on peanut stem. **b.** Section through perithecium with asci. **c.** Ascus with ascospores. **d.** Mature ascospore.

GIBBERELLA Sacc.

Ascoma an ostiolate perithecium, borne free or associated with a stroma; stroma, if present, erumpent and pseudoparenchymatous or effuse. Perithecia superficial to partially immersed, scattered to caespitose, fleshy, ovoid to globose, with an ostiolar papilla, blue to violet. Ascomal wall pseudoparenchymatous. Centrum with apical paraphyses. Asci unitunicate, with an undifferentiated apex, nonamyloid, ellipsoid to broadly clavate, 8-spored. Ascospores phragmosporous, but often variable in septation, ellipsoid, fusoid or cymbiform, straight or curved, smooth, hyaline to slightly yellowish.

Anamorph: *Fusarium.*

Habitat: On stems or wood.

Representative species: *Gibberella fujikuroi* (Sawada) Ito (Anam. *Fusarium moniliforme* J. Sheld.), cause of bakanae disease of rice (*Oryza* spp.) and source of gibberellins.

Comments: In *Calonectria* De Not. and *Nectria* (Fr.) Fr. the perithecia are brightly colored; in *Micronectriella* Höhn. the ascomata are immersed in host tissue.

References: Booth, 1971; Dennis, 1981; Dingley, 1952b; Domsch et al., 1980; Ellis and Ellis, 1985; Ellis and Everhart, 1892; CMI 22-24, 310, 384-385, 574.

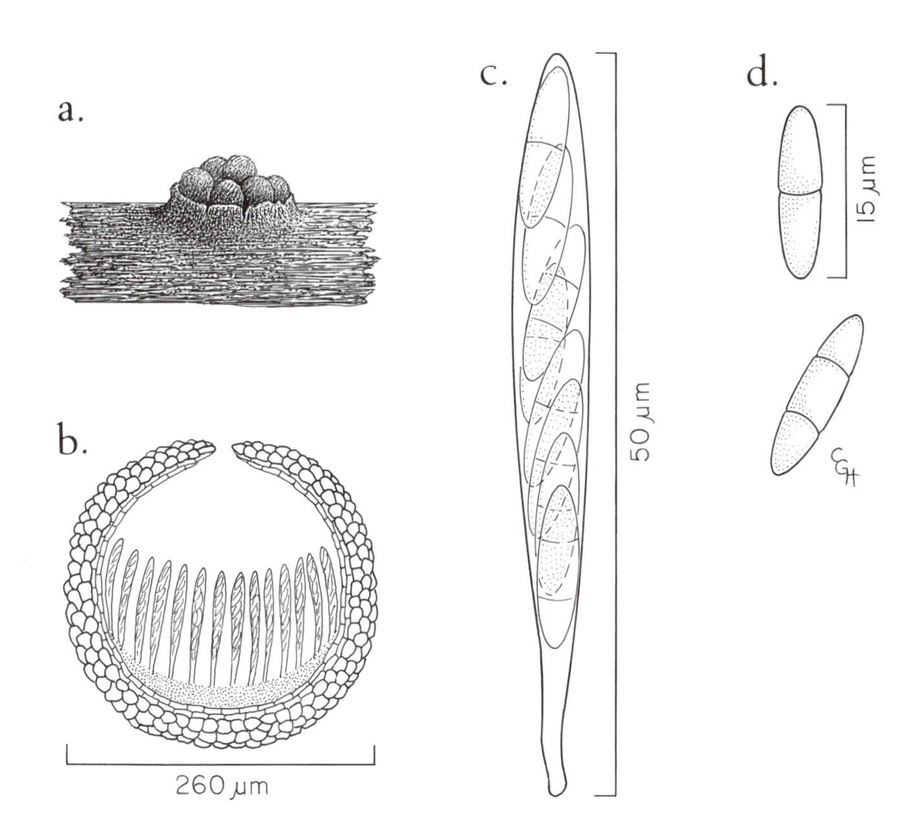

Gibberella zeae. **a.** Cluster of perithecia on erumpent stroma on stem of maize. **b.** Section through perithecium with asci. **c.** Ascus with ascospores. **d.** Mature ascospores.

SETOSPHAERIA K. J. Leonard & E. G. Suggs

Ascoma a uniloculate, perithecioid pseudothecium, superficial, erumpent or immersed in host tissues; pseudothecia dark brown to black, ostiolate, with or without a neck, with short brown setae around ostiole and on upper part of pseudothecium. Ascomal wall composed of angular to globose pseudoparenchymatous cells, coriaceous or carbonaceous. Centrum with filiform, septate, branched pseudoparaphyses. Asci bitunicate, thick-walled, cylindrical or clavate, straight or curved, stalked, 1-8-spored. Ascospores hyaline, fusoid to oblong, several septate, constricted at septa, smooth, surrounded by a thin gelatinous sheath.

Anamorph: *Exserohilum* (*Drechslera*).

Habitat: Mostly parasitic on grasses or other monocotyledons, but some species also found in soil and other substrates.

Representative species: *Setosphaeria turcica* (Luttrell) K. J. Leonard & E. G. Suggs [Anam. *Exserohilum turcicum* (Pass.) K. J. Leonard and E. G. Suggs =*Drechslera turcica* (Pass.) Subramanian & P. C. Jain], cause of leafspot disease of sorghum (*Sorghum* spp.) and maize (*Zea mays* L.).

Comments: *Keissleriella* Höhn. differs in its superficial ascomata, the presence of a clypeus, the presence of setae on the pseudothecial wall, the lack of periphyses, and different anamorphs. *Trichometasphaeria* Munk is regarded as synonymous with *Keissleriella.*

References: Leonard and Suggs, 1974; Sivanesan, 1984; CMI 304, 587.

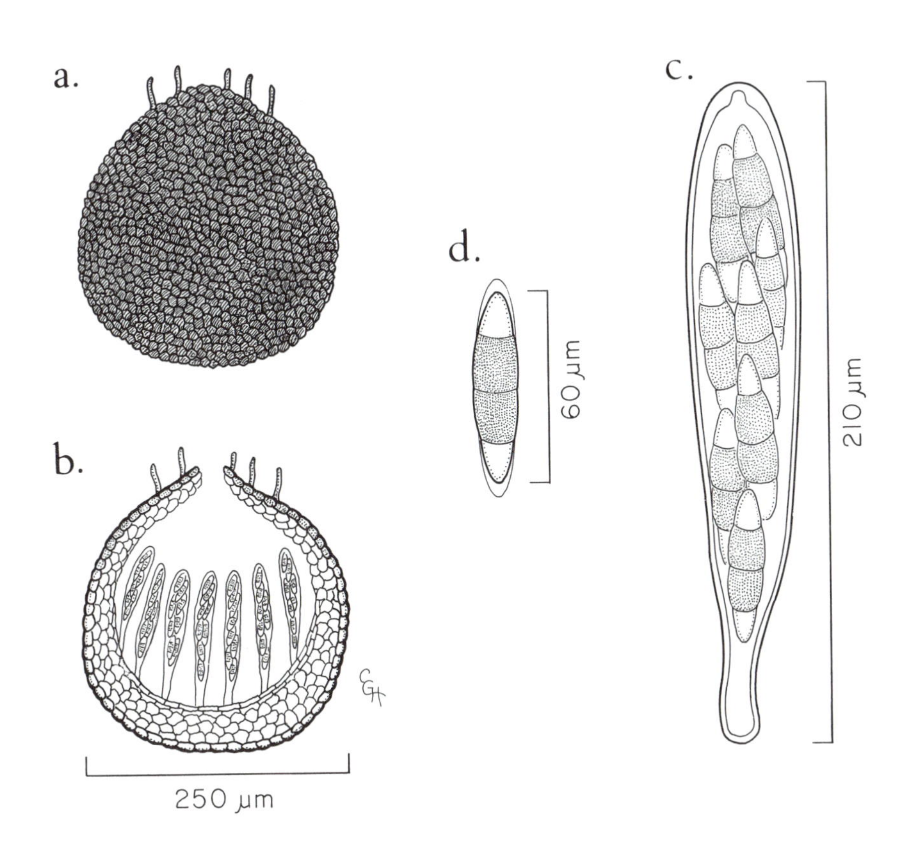

Setosphaeria turcica. **a.** Exterior of pseudothecium. **b.** Section through pseudothecium with asci. **c.** Ascus with ascospores. **d.** Mature ascospores.

SPHAERULINA Sacc.

Ascoma a uniloculate, perithecioid pseudothecium, scattered or grouped, sometimes confluent and appearing multiloculate, immersed or erumpent; pseudothecia brown, globose or conical, ostiolate, with an ostiolar papilla. Ascomal wall composed of brown, polygonal cells, thick- or thin-walled. Asci bitunicate, oblong, clavate or cylindric-clavate, fasciculate, stalked or sessile, 8-spored. Ascospores phragmosporous, cylindric to filiform, straight or curved, guttulate, smooth, hyaline, but sometimes becoming light brown upon discharge.

Anamorphs: *Cercospora, Cercosporella,* and *Septoria.*

Habitat: On leaves of dicotyledonous plants.

Representative species: *Sphaerulina oryzina* K. Hara (Anam. *Cercospora oryzae* Miyake), cause of leafspot of rice.

Comments: *Sphaerulina* differs from *Mycosphaerella* Johans. in having phragmosporous ascospores.

References: Barr, 1972; Dennis, 1981; Ellis and Ellis, 1985; Ellis and Everhart, 1892; Munk, 1957; Sivanesan, 1984; IM 23:C348.

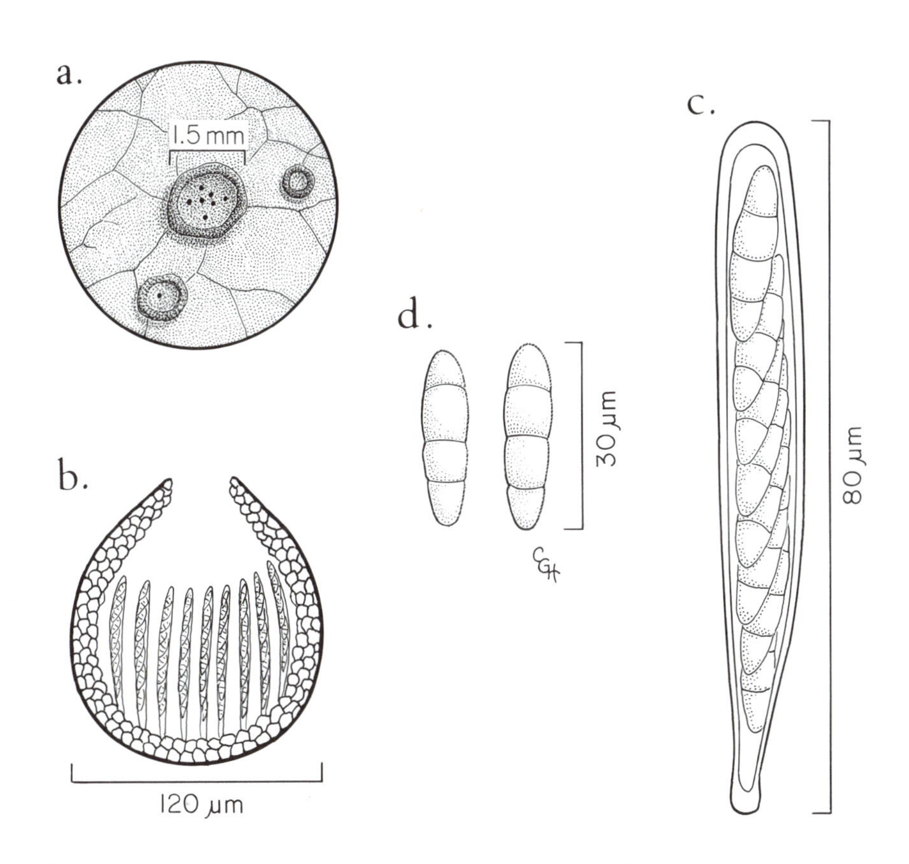

Sphaerulina plantiginea. **a.** Lesion with ostiolar necks on leaf of *Plantago*. **b.** Section through pseudothecium with asci. **c.** Ascus with ascospores. **d.** Mature ascospores.

SCORIAS Fr.:Fr.

Mycelium forming a black subiculum over surface of host; subiculum in some species large and sponge-like and composed of branching bundles of united hyphae; pseudothecia and pycnidia borne on surface of subiculum, the entire mass covered by a thin gelatinous sheath. Pseudothecia black, borne singly, but often crowded, subglobose to ovate or broadly ellipsoid, with an elongate sterile base, glabrous. Asci bitunicate, oblong to saccate. Ascospores phragmosporous, subclavate to clavate, hyaline, but often becoming olivaceous, smooth or finely roughened at maturity.

Anamorphs: *Conidiocarpus, Polychaeton,* and *Scolecoxyphium.*

Habitat: On branches of trees and shrubs.

Representative species: *Scorias spongiosa* (Schwein.:Fr.) Fr. (Anam. *Polychaeton* sp.), on leaves and branches of beech (*Fagus* sp.).

Comments: In *Trichomerium* Speg. the ascomata are setose and a subiculum is lacking. Hughes regards *Trichomerium* as a synonym of *Phragmocapnias* Theiss. & Syd.

References: Barr, 1972; Ellis and Everhart, 1892; Hughes, 1976; Sivanesan, 1984.

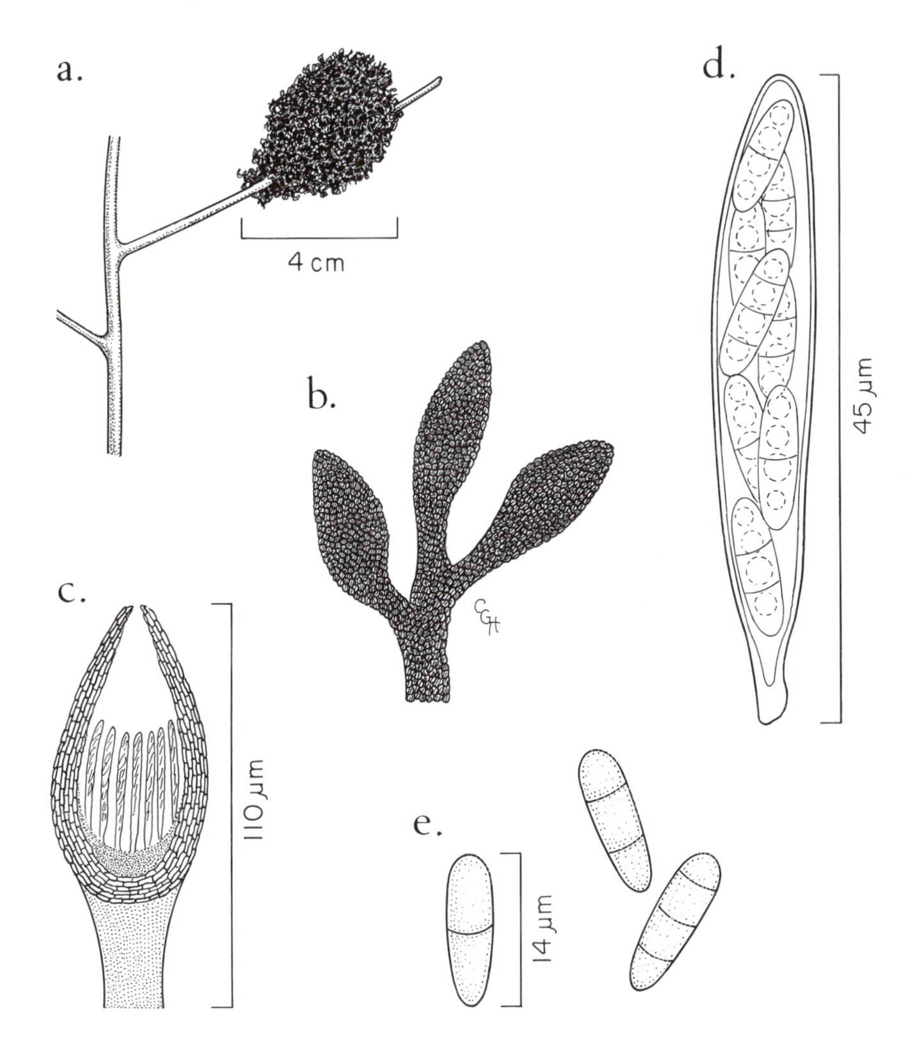

Scorias spongiosa. **a.** Mycelial subiculum on beech twig. **b.** Exterior of ascigerous locules on mycelium. **c.** Section through ascigerous locule with asci. **d.** Ascus with ascospores. **e.** Mature ascospores.

ELSINOE **RACIB.**

Mycelium forming scab-like lesions on host tissues, variable in shape, hyaline to pale brown. Ascomata formed in lesions, separate or aggregated, epidermal or subepidermal and partially erumpent, globose, circular, elliptical or pulvinate, composed of angular pseudoparenchyma cells, hyaline to pale brown. Asci distributed irregularly in ascomal tissue in uniascal locules, bitunicate, ovoid, globose, subglobose or elliptical, sessile, thick-walled, 8-spored. Ascospores phragmosporous, but sometimes with a vertical septum, hyaline to pale yellow, usually constricted at middle septum.

Anamorph: *Sphaceloma.*

Habitat: Parasitic on leaves, stems, scale insects, or other fungi.

Representative species: *Elsinoë fawcettii* Bitancourt & Jenk. (Anam. *Sphaceloma fawcettii* Jenk.), cause of scab disease of *Citrus* spp.

Comments: In *Bitancourtia* Thirum. & Jenk. the ascoma is fully erumpent and the ascospores have occasional longitudinal septa. Von Arx and Müller regard this as a synonym of *Elsinoë.*

References: Arx, 1963; Dennis, 1981; Sivanesan, 1984; Viégas, 1944; CMI 313-314, 439-440, 484; IM 5:C53.

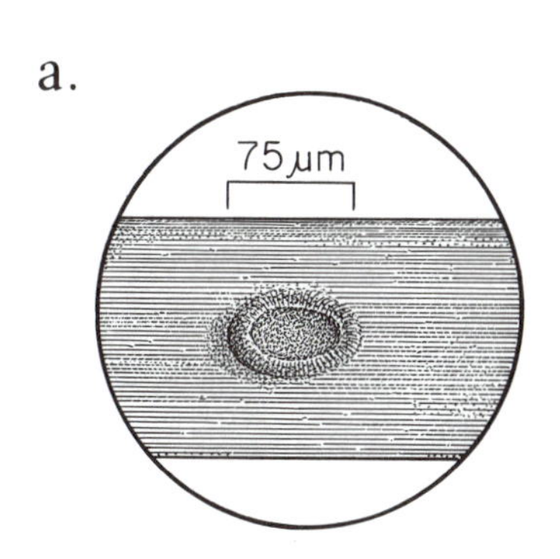

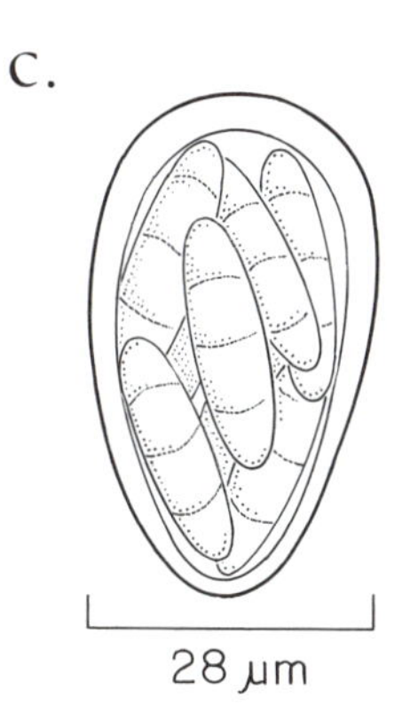

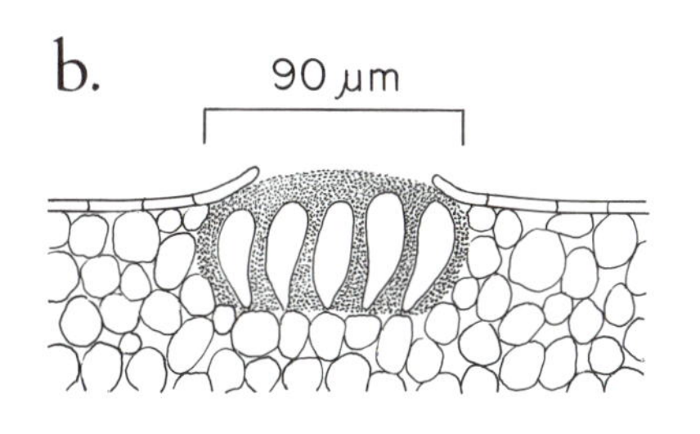

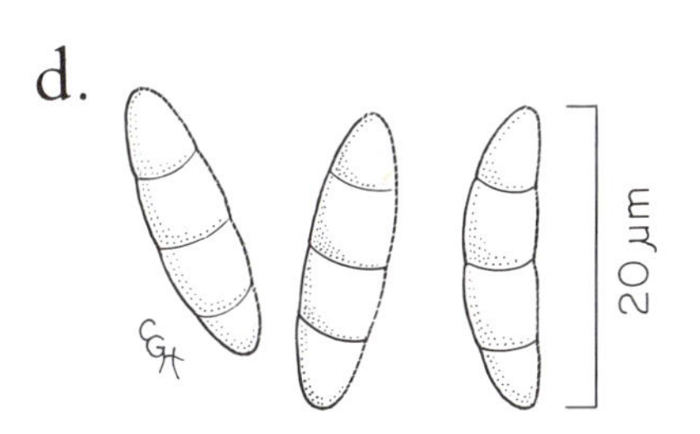

Elsinoë veneta. **a.** Erumpent ascostroma on stem of *Rubus.* **b.** Section through ascostroma with asci. **c.** Ascus with ascospores. **d.** Mature ascospores.

MELIOLA **Fr.**

Ascoma a perithecium borne on surface of mycelium; mycelium superficial on host, dark brown, much-branched, bearing erect, dark brown setae, forming capitate and mucronate hyphopodia. Perithecia single, globose, ostiolate, but without an ostiolar neck. Ascomal wall composed of several layers of pseudoparenchyma cells, outer cells thick-walled and dark brown, inner cells thin-walled, subhyaline and flattened. Asci unitunicate, formed in a basal fascicle, slightly thickened at apex, 2-8-spored. Ascospores phragmosporous, dark brown, cylindric, ellipsoid or subfusoid, oblong, straight or curved, smooth.

Anamorph: None, but Hughes reported observing conidia on mucronate hyphopodia in one species.

Habitat: Parasitic on living leaves and young stems of flowering plants.

Representative species: *Meliola circinans* Earle.

Comments: *Appendiculella* Höhn., *Asteridiella* McAlpine, and *Irenopsis* F. Stevens all differ from *Meliola* in the lack of mycelial setae; in *Irenopsis* the perithecia bear setae, in *Appendiculella* they bear larviform appendages, and in *Asteridiella* the perithecia are glabrous.

References: Dennis, 1981; Ellis and Ellis, 1985; Ellis and Everhart, 1892; Hansford, 1961, 1963; Viégas, 1944.

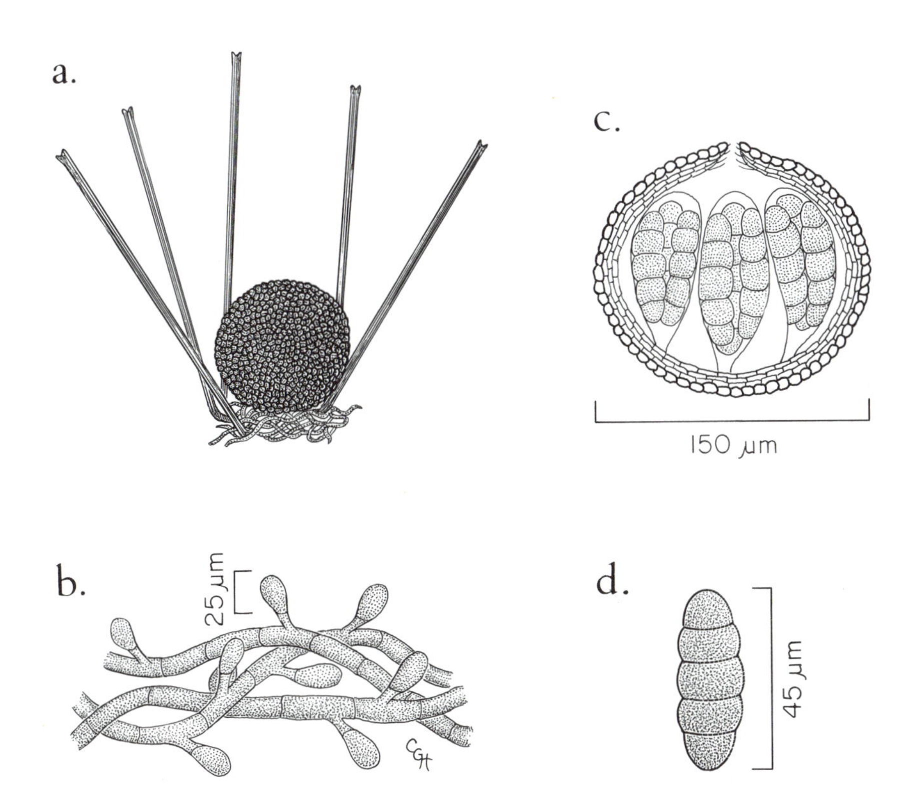

Meliola bidentata. **a.** Perithecium surrounded by mycelial setae. **b.** Close-up of hyphae with lateral capitate hyphopodia. **c.** Section through perithecium with asci. **d.** Mature ascospore.

LEPTOSPHAERIA Ces. & De Not.

Ascoma a uniloculate, perithecioid pseudothecium, scattered or aggregated, immersed in host tissue, but sometimes exposed at maturity, ostiolate, dark brown, glabrous. Ostiolar neck short or lacking, with or without periphyses. Ascomal wall several cells thick, composed of isodiametric, thick-walled scleroplectenchyma, especially near base of ostiolar neck. Centrum containing broad, septate pseudoparaphyses. Asci bitunicate, cylindric to clavate, short-stalked, 4-8-spored. Ascospores phragmosporous, fusiform, cylindric or clavate, often constricted at median septum, supramedian cell often broader than others, yellowish to brown, smooth or rarely echinulate, often with terminal globose appendages, sometimes covered by gelatinous sheath.

Anamorphs: *Coniothyrium, Diplodina, Phaeoseptoria, Phoma, Scolecosporella, Septoria*, and *Stagonospora.*

Habitat: On stems of herbaceous dicots.

Representative species: *Leptosphaeria sacchari* Breda de Hahn [Anam. *Phoma* sp. (as *Phyllosticta*)], cause of ring spot of sugar cane (*Saccharum* spp.).

Comments: *Phaeosphaeria* Miyake has been used for species on monocotyledons, but is regarded as a synonym of *Leptosphaeria* by most authors. In *Paraphaeosphaeria* O. Eriksson the first septum is submedian; in *Nodulosphaeria* Rabenh. it is supramedian.

References: Dennis, 1981; Domsch et al., 1980; Ellis and Ellis, 1985; Ellis and Everhart, 1892; Eriksson, 1967b; Holm, 1952, 1957; Koponen and Makela, 1975; Matsushima, 1971; Müller, 1950; Munk, 1957; Shoemaker, 1984; Sivanesan, 1984; BK 381-382; CMI 331, 506, 663, 771; IM 3:C33.

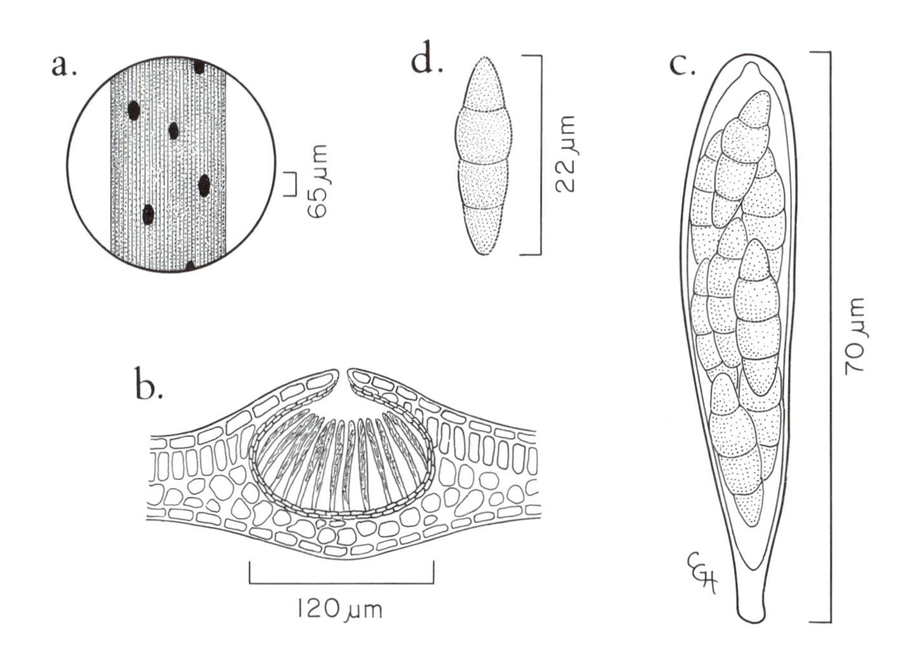

Leptosphaeria sacchari. **a.** Lesions on leaf of sugar cane. **b.** Section through pseudothecium with asci. **c.** Ascus with ascospores. **d.** Mature ascospore.

SPORORMIELLA Ellis & Everh.

Ascoma a uniloculate, perithecioid pseudothecium, scattered or aggregated, partly or completely immersed in substrate, dark brown to black, smooth or with hairs on upper part, ostiolate; ostiolar neck papillate or elongated. Ascomal wall membranaceous to coriaceous, composed of pseudoparenchyma cells, outer cells somewhat thick-walled and pigmented, inner cells thin-walled and hyaline. Centrum containing filamentous pseudoparaphyses. Asci bitunicate, clavate to cylindrical, with short or long stalk, 8-spored. Ascospores dark brown, phragmosporous, strongly constricted at septa, septa transverse or oblique, with a germ slit; germ slit diagonal, transverse or parallel to long axis of spore; spore surrounded by a gelatinous sheath.

Anamorph: None reported.

Habitat: On dung of various animals.

Representative species: *Sporormiella leporina* (Niessl) Ahmed & Cain (=*Sporormia leporina* Niessl) on various types of dung.

Comments: In *Sporormia* De Not. the ascospores are joined together in a common gelatinous sheath, and in *Preussia* Fuckel the ascomata are nonostiolate.

References: Ahmed and Cain, 1972; Dennis, 1981; Matsushima, 1975; Munk, 1957; IM 9:C119.

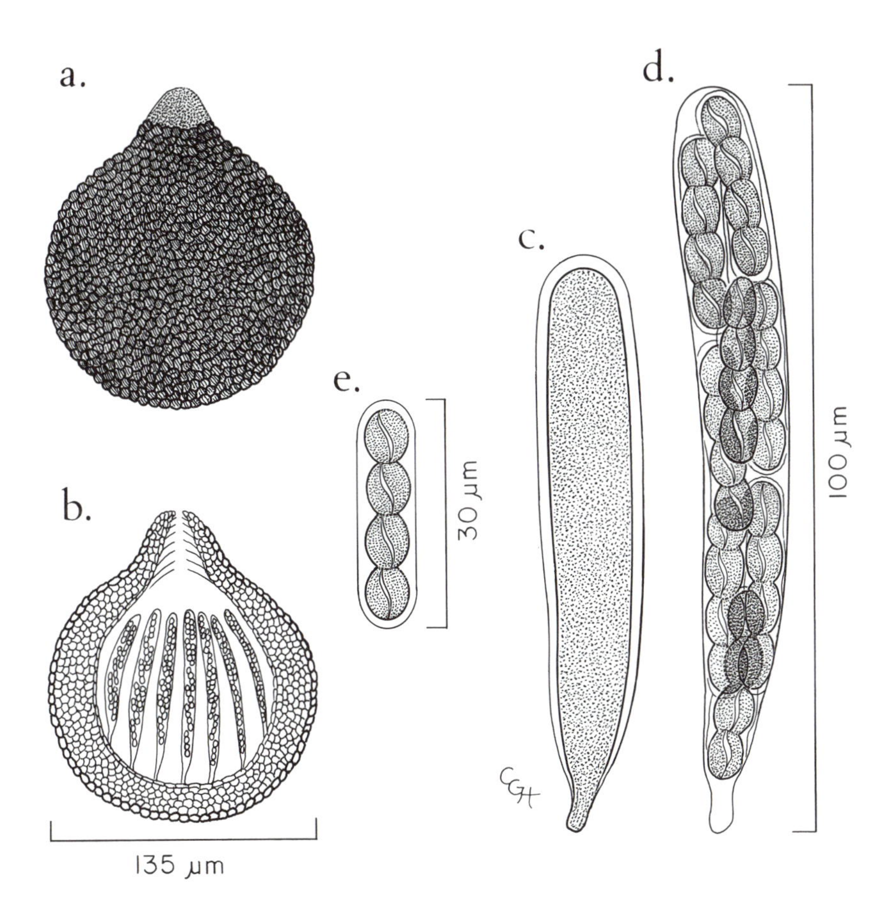

Sporormiella minima. **a.** Exterior of pseudothecium. **b.** Section through pseudothecium with asci. **c.** Young bitunicate ascus. **d.** Ascus with ascospores. **e.** Mature ascospore with germ slits and gelatinous sheath.

HYSTERIUM Pers.:Fr.

Ascoma an elongate hysterothecium, single, superficial, black, dull, carbonaceous, oval to elongated or linear, straight or curved, rarely branched, opening by a sunken longitudinal slit. Centrum containing pseudoparaphyses. Asci bitunicate, clavate or cylindrical, 8-spored. Ascospores phragmosporous, spindle-shaped, oblong, elliptical or cylindrical, brown.

Anamorph: *Coniosporium.*

Habitat: On dead wood of vascular plants.

Representative species: *Hysterium pulicare* Pers.:Fr., occurring on a wide range of hosts.

Comments: In *Gloniella* Sacc. the phragmosporous ascospores are hyaline, and in *Hysterographium* Corda emend. De Not. the ascospores are brown and dictyosporous.

References: Dennis, 1981; Zogg, 1962; BK 387; IM 18:C256.

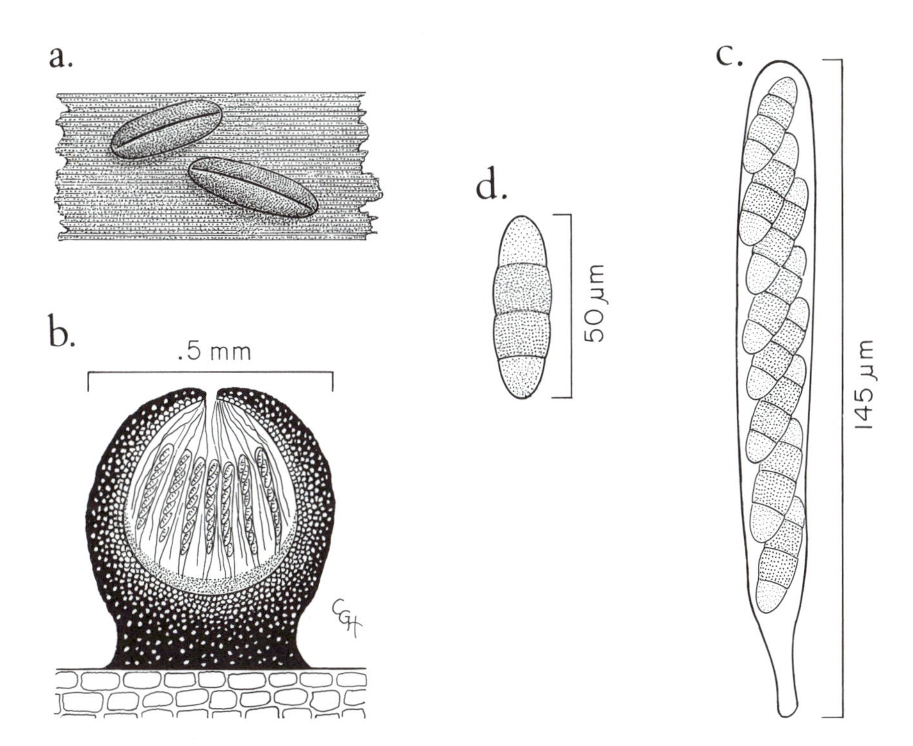

Hysterium pulicare. **a.** Two hysterothecia on wood. **b.** Section through hysterothecium with asci. **c.** Ascus with ascospores. **d.** Mature ascospore.

GEOGLOSSUM Pers.

Ascomata erect, clavate-stipitate, brown or black, fleshy, dry, viscid or gelatinous; clavate upper portion of ascoma laterally flattened and bearing hymenium; stipe slender, terete, smooth, pubescent, squamulose or setose-hirsute. Paraphyses straight or curved, often enlarged at apices, intermixed with the asci, in some species extending down stipe as tufts or as a continuous gelatinous layer. Asci unitunicate, inoperculate, clavate, 2-8-spored. Ascospores clavate, subcylindric or subfusoid, often tapering toward the lower end, usually parallel in ascus, brown or hyaline, or a mixture of both, phragmosporous, but occasionally nonseptate.

Anamorph: None reported.

Habitat: On soil and in rotting wood.

Representative species: *Geoglossum simile* Peck, common in swamps and bogs and on humus and rotten wood.

Comments: Species of *Geoglossum* with a gelatinous layer of paraphyses on the stipe have been placed in *Gloeoglossum* Durand, but this is considered a synonym of *Geoglossum. Trichoglossum* Boud. differs in having abundant dark brown, acuminate setae in the hymenium and on the stipe.

References: Dennis, 1981; Mains, 1954; BK 132-133.

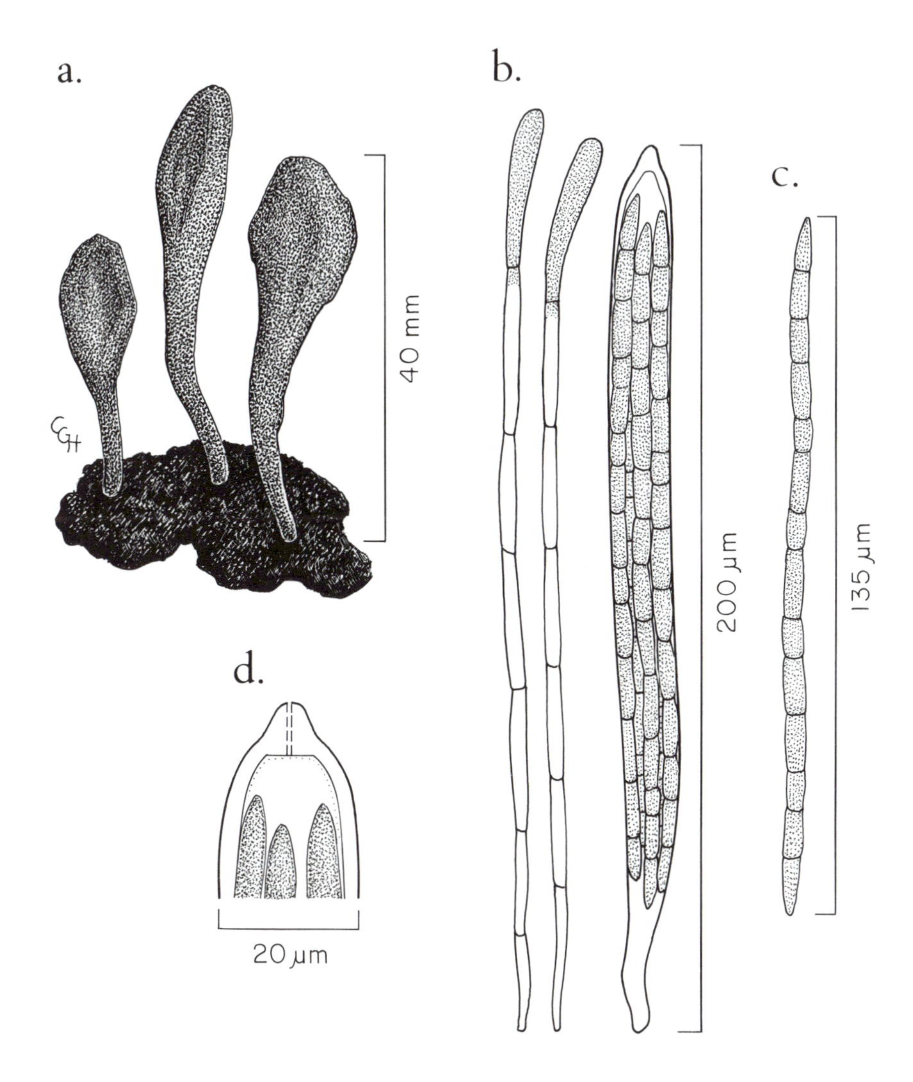

Geoglossum difforme. **a.** Ascomata on soil. **b.** Paraphyses and ascus with ascospores. **c.** Mature ascospore. **d.** Ascus apex.

MYRIANGIUM Mont. & Berk.

Ascoma a multiloculate ascostroma formed on a basal stroma growing on the substrate; ascostroma black, cushion-shaped to shallow discoid or cup-shaped, composed of fertile and sterile portions, pseudoparenchymatous. Asci bitunicate, globose or broadly ellipsoid, scattered at different levels in uniascal locules, becoming free through disintegration of stroma. Ascospores dictyosporous, oblong to oval or elliptical, hyaline or yellowish.

Anamorph: None reported.

Habitat: On leaves and stems of living plants, usually parasitic on scale insects.

Representative species: *Myriangium duriaei* Mont. & Berk.

Comments: In *Butleria* Sacc. the ascostroma is light-colored and the ascospores are 2-celled; in *Angatia* Syd. the muriform ascospores are formed in broadly clavate asci borne in a layer in the ascostroma, and in *Anhellia* Racib. the cushion-shaped ascomata are erumpent from host tissues.

References: Arx, 1963; Dennis, 1981; Ellis and Ellis, 1985; Miller, 1940; Viégas, 1944; IM 5:C52.

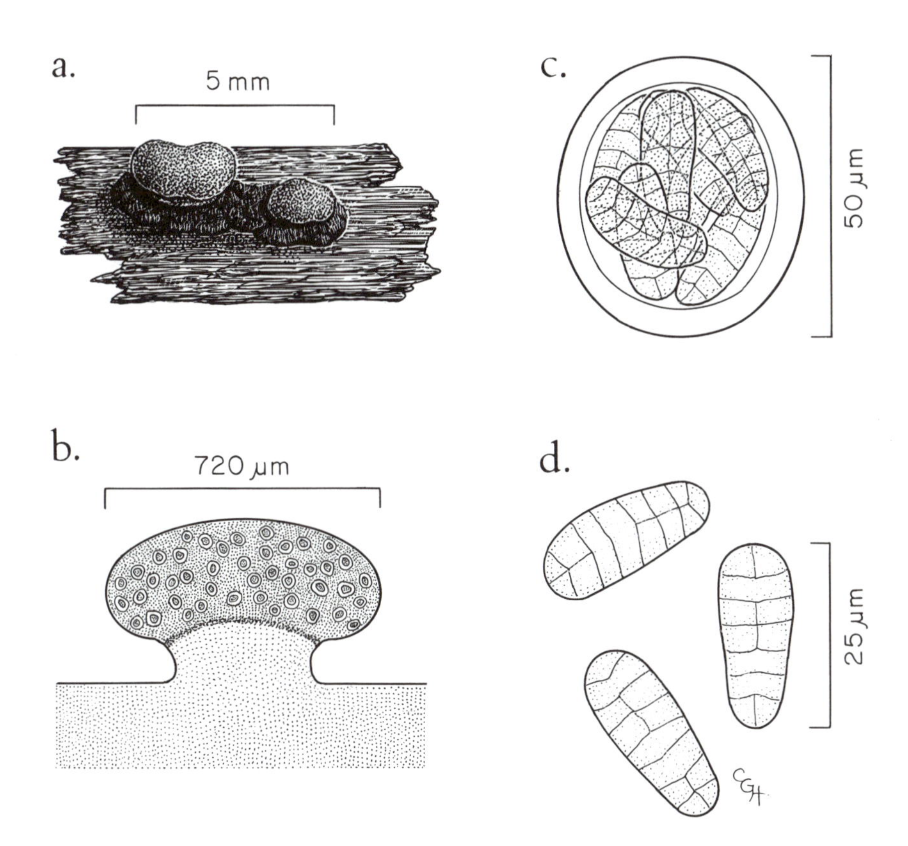

Myriangium durieae. **a.** Ascostromata on bark. **b.** Section through ascostroma with fertile region and steile base. **c.** Ascus with ascospores. **d.** Mature ascospores.

LEPTOSPHAERULINA McAlpine

Ascoma an ostiolate, uniloculate, perithecioid pseudothecium, immersed in host tissue, apex erumpent at maturity. Ascoma composed of pseudoparenchyma cells, outer cells brown and thick-walled, interior cells hyaline and thin-walled. Centrum pseudoparenchymatous. Asci few, bitunicate, saccate, thick-walled, 8-spored. Ascospores hyaline, variable in shape from oblong to ellipsoid or short cylindric, dictyosporous, but longitudinal septa sometimes lacking in some spores, with a thin gelatinous sheath, sometimes becoming brownish with age.

Anamorph: *Pithomyces.*

Habitat: On leaves and herbaceous stems.

Representative species: *Leptosphaerulina crassiasca* (Sechet) C. R. Jackson & D. K. Bell, cause of leaf scorch and pepper spot of peanut.

Comments: The genus *Pseudoplea* Höhn. is considered a synonym of *Leptosphaerulina. Pleospora* Rabenh. ex Ces. & De Not. differs in its distinctly brown ascospores and in the presence of pseudoparaphyses.

References: Barr, 1972; Dennis, 1981; Ellis and Ellis, 1985; Graham and Luttrell, 1961; Irwin and Davis, 1985; Matsushima, 1971; Viégas, 1944; CMI 146; IM 38:C746.

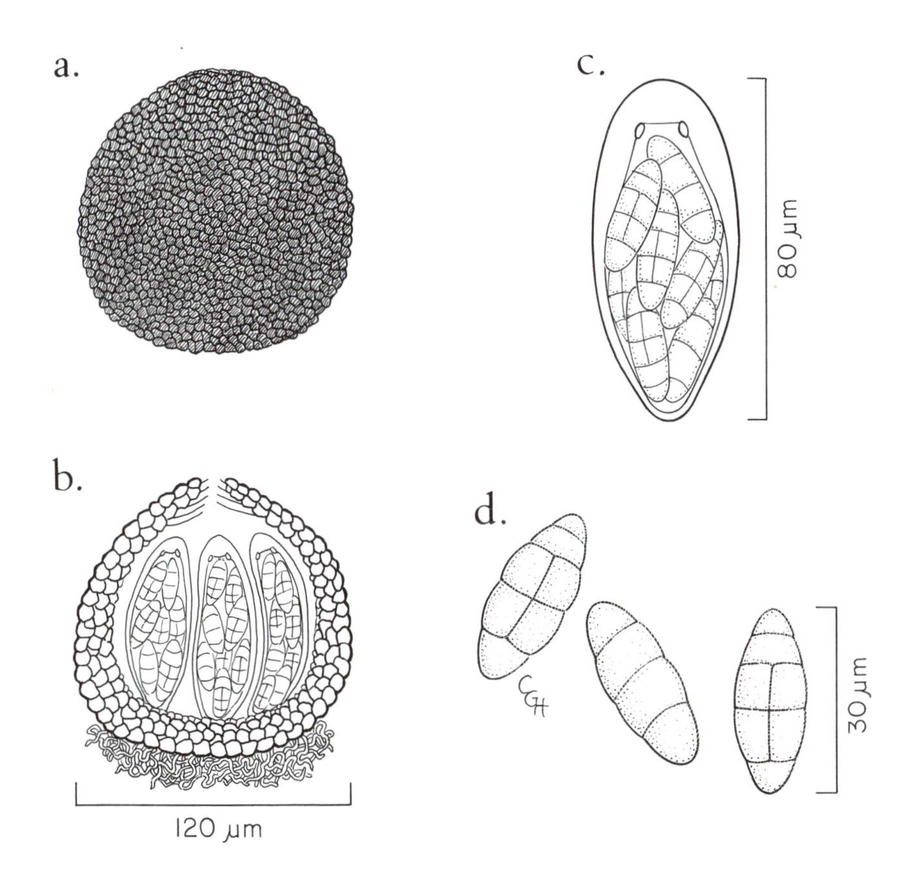

Leptosphaerulina crassiasca. **a.** Exterior of pseudothecium. **b.** Section through pseudothecium with asci. **c.** Ascus with ascospores. **d.** Mature ascospores.

PLEOSPORA Rabenh. ex Ces. & De Not.

Ascoma a uniloculate, perithecioid pseudothecium, scattered and immersed in host tissues; pseudothecia brown to black, ostiolate, with a short ostiolar papilla. Ascomal wall composed of pseudoparenchyma cells, outer cells angular, thick-walled and dark brown, inner cells hyaline, smaller and flattened on inside of wall. Centrum containing numerous filiform, branched pseudoparaphyses. Asci bitunicate, cylindric-clavate, short-stalked, 6-8-spored. Ascospores dictyosporous, ellipsoidal, oblong to clavate, overlapping in ascus, smooth or verrucose, yellow-brown to dark brown, sometimes with a gelatinous sheath.

Anamorphs: *Alternaria, Camarosporium, Coniothyrium, Diplodia, Hendersonia, Microdiplodia, Phoma,* and *Stemphyllium.*

Habitat: Mostly occurring on leaves and stems of herbaceous plants, with a few species occurring on woody plants.

Representative species: *Pleospora herbarum* (Pers.:Fr.) Rabenh. (Anam. *Stemphyllium herbarum* E. Simmons), on a wide range of host plants.

Comments: *Clathrospora* Rabenh. and *Platyspora* Wehmeyer differ in having dictyosporous ascospores that are distinctly flattened. In *Platyspora* the vertical septa are continuous, whereas in *Clathrospora* they are irregularly placed. Simmons restricts *Pleospora* to species with *Stemphyllium* anamorphs and places those species with *Alternaria* anamorphs in the genus *Lewia* Barr & E. Simmons.

References: Dennis, 1981; Domsch et al., 1980; Ellis and Ellis, 1985; Ellis and Everhart, 1892; Eriksson, 1967; Matsushima, 1975; Müller, 1951; Simmons, 1986; Sivanesan, 1984; Viégas, 1944; Wehmeyer, 1961; CMI 149-150, 730; FC 232; IM 1:C3.

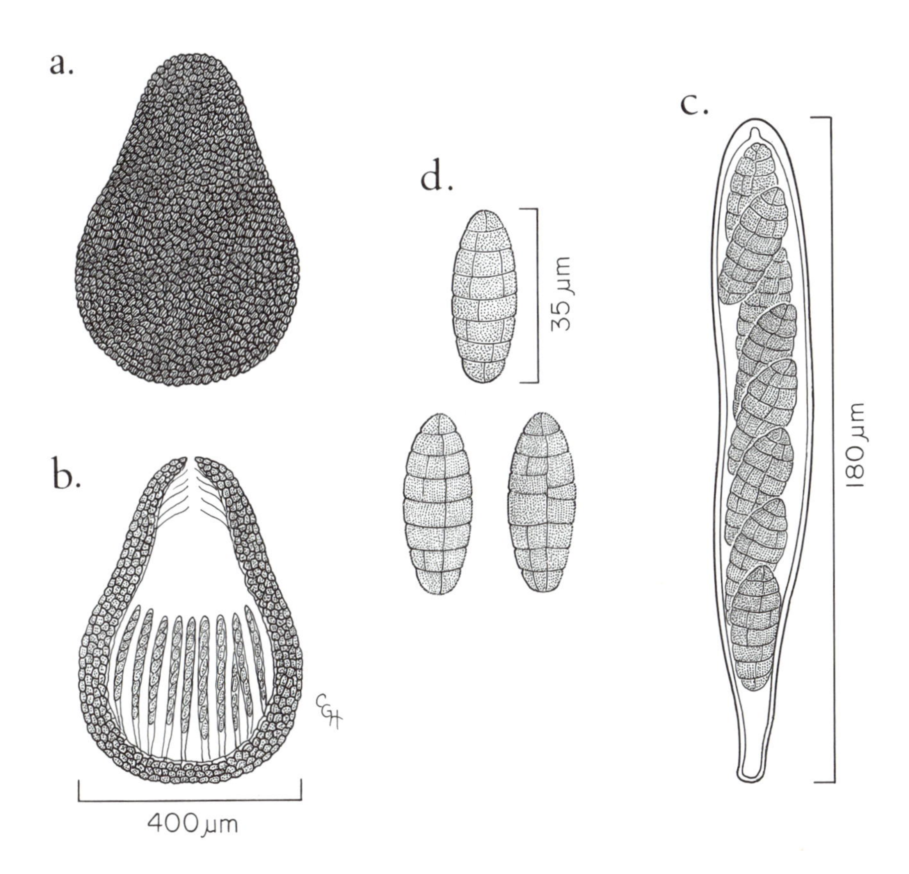

Pleospora herbarum. **a.** Exterior of pseudothecium. **b.** Section through pseudothecium with asci. **c.** Ascus with ascospores. **d.** Mature ascospore.

PYRENOPHORA Fr.

Ascoma a uniloculate, perithecioid pseudothecium, solitary and scattered, immersed in host tissues, erumpent, sometimes nearly superficial. Pseudothecia black, globose to somewhat flattened, ellipsoidal or obpyriform, ostiolate, with a short papilla, bearing dark brown setae on upper surface. Ascomal wall pseudoparenchymatous, outer cells dark brown and thick-walled, becoming hyaline and thin-walled on the inside. Centrum containing indefinite, filiform, branched and anastomosing pseudoparaphyses. Asci bitunicate, thick-walled, broadly clavate to saccate, short-stalked, with a nonamyloid apical ring, 2-8-spored. Ascospores dictyosporous, irregularly biseriate in ascus, slightly constricted at septa, hyaline to pale yellow or yellow-brown, smooth, or slightly verrucose, with or without a gelatinous sheath.

Anamorph: *Drechslera.*

Habitat: On leaves and stems of herbaceous plants, often parasitic.

Representative species: *Pyrenophora avenae* Ito & Kuribayashi [Anam. *Dreschlera avenae* (Eidam) Scharif], cause of eye spot disease of oats (*Avena* spp.) and other grasses.

Comments: *Pleospora* Rabenh. ex Ces. & De Not. differs in having ascospores that are dark brown and usually smaller, and in the anamorph.

References: Dennis, 1981; Ellis and Ellis, 1985; Sivanesan, 1984; Wehmeyer, 1961; CMI 388-390, 493-494; IM 9:C127.

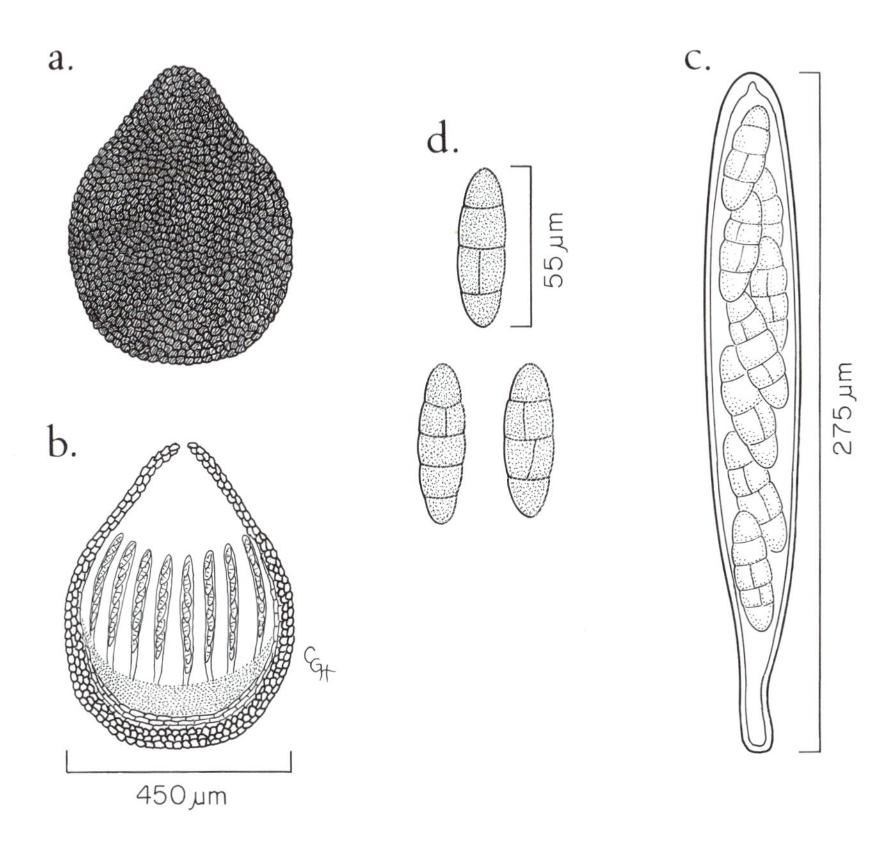

Pyrenophora avenae. **a.** Exterior of pseudothecium. **b.** Section through pseudothecium with asci. **c.** Ascus with ascospores. **d.** Mature ascospores.

HYSTEROGRAPHIUM Corda emend. De Not.

Ascoma an elongate hysterothecium, single, superficial, black, dull, carbonaceous, oval to elongate or linear, straight or curved, rarely branched, opening by a sunken longitudinal slit. Centrum containing pseudoparaphyses. Asci bitunicate, saccate, clavate or cylindrical, 8-spored. Ascospores dictyosporous, spindle-shaped, elliptical, oval, oblong to cylindrical, brown.

Anamorph: *Hysteropycnis.*

Habitat: On dead wood.

Representative species: *Hysterographium fraxini* (Pers.:Fr.) De Not.

Comments: In *Gloniopsis* De Not. the dictyosporous ascospores are hyaline.

References: Dennis, 1981; Ellis and Ellis, 1985; Zogg, 1962; IM 18:C255.

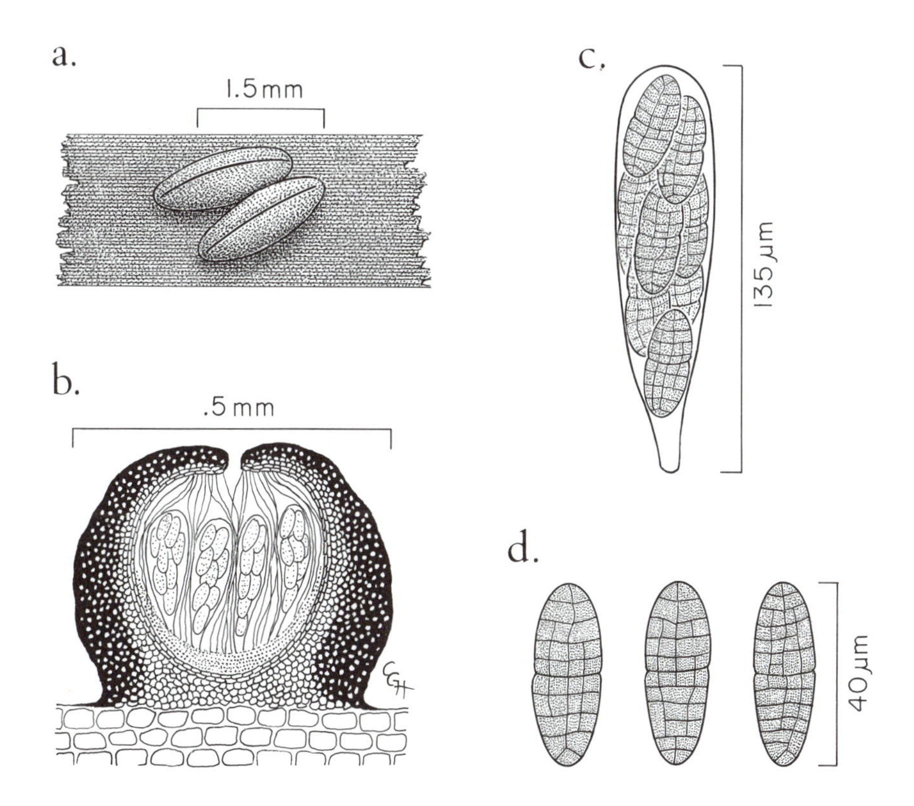

Hysterographium fraxini. **a.** Hysterothecia on wood. **b.** Section through hysterothecium with asci. **c.** Ascus with ascospores. **d.** Mature ascospores.

REFERENCES

Ahmed, S. I., and R. F. Cain. 1972. Revision of the genera *Sporormia* and *Sporormiella.* Can. J. Bot. 50:419-477.

Alcorn, J. L. 1982. New *Cochliobolus* and *Bipolaris* species. Mycotaxon 15:1-19.

Alcorn, J. L. 1983. On the genera *Cochliobolus* and *Pseudocochliobolus.* Mycotaxon 16:353-379.

Ames, L. M. 1961. A monograph of the Chaetomiaceae. U.S. Army Res. Dev. Ser., No.2:1-125.

Arnaud, G. 1918. Les Astérinées. Ann. École Nat. Agric. Montpellier N.S. 16:1-288.

Arx, J. A. von. 1949. Beiträge zur Kenntnis der Gattung *Mycosphaerella.* Sydowia 3:28-100.

Arx, J. A. von. 1961. Über *Cylindrosporium padi*. Phytopathol. Z. 42:161-166.

Arx, J. A. von. 1963. Die Gattungen der Myriangiales. Persoonia 2:421-475.

Arx, J. A. von. 1975a. On *Thielavia* and some similar genera of Ascomycetes. Stud. Mycol. 8:1-31.

Arx, J. A. von. 1975b. Revision of *Microascus* with the description of a new species. Persoonia 8:191-197.

Arx, J. A. von. 1982. A key to the species of *Gelasinospora.* Persoonia 11:443-449.

Arx, J. A. von, M. Dreyfuss, and E. Müller. 1984. A revaluation of *Chaetomium* and the Chaetomiaceae. Persoonia 12:169-179.

Arx, J. A. von, J. Guarro, and M. J. Figueras. 1986. The ascomycete genus *Chaetomium.* Nova Hedw. Suppl. 84:1-162.

Arx, J. A. von, and E. Müller. 1954. Die Gattungen der amerosporen Pyrenomyceten. Beitr. Kryptogamenfl. Schweiz 11(1):1-434.

Barr, M. E. 1968. The Venturiaceae in North America. Can. J. Bot. 46:799-864.

Barr, M. E. 1970. Some amerosporous ascomycetes on Ericaceae and Empetraceae. Mycologia 62:377-394.

Barr, M. E. 1972. Preliminary studies on the Dothideales in temperate North America. Contrib. Univ. Mich. Herb. 9:523-638.
Barr, M. E. 1978. The Diaporthales in North America. Mycol. Mem. 7:1-232.
Barron, G. L., R. F. Cain, and J. C. Gilman. 1961. The genus *Microascus.* Can. J. Bot. 39:1609-1631.
Batra, L. R. 1960. The species of *Ciborinia* pathogenic to *Salix, Magnolia,* and *Quercus.* Am. J. Bot. 47:819-817.
Batra, L. R. 1983. *Monilinia vaccinii-corymbosi* (Sclerotiniaceae): its biology on blueberry and comparison with related species. Mycologia 75:131-152.
Batra, L. R., and Y. Harada. 1986. A field record of apothecia of *Monilinia fructigena* in Japan and its significance. Mycologia 78:913-917.
Batra, L. R., and R. P. Korf. 1959. The species of *Ciborinia* pathogenic to herbaceous angiosperms. Am. J. Bot. 46:441-450.
Bell, D. K., and E. K. Sobers. 1966. A peg, pod, and root necrosis of peanuts caused by a species of *Calonectria.* Phytopathology 56:1361-1364.
Benjamin, R. K. 1949. Two new species representing a new genus of the Chaetomiaceae. Mycologia 41:347-354.
Bissett, J. 1986a. A note on the typification of *Guignardia.* Mycotaxon 25:519-522.
Bissett, J. 1986b. *Discochora yuccae* sp. nov. with *Phyllosticta* and *Leptodothiorella* synanamorphs. Can. J. Bot. 64:1720-1726.
BK. Breitenbach, J., and F. Kränzlin. 1984. Fungi of Switzerland. Volume 1. Ascomycetes. Verlag Mykologia, Luzern. 310 pp. (Also available in French and German editions. References are to photograph numbers).
Blumer, S. 1967. Echte Meltaupilze (Erysiphaceae). Gustav Fischer Verlag, Jena. 436 pp.
Booth, C. 1959. Studies of Pyrenomycetes: IV. *Nectria* (Part I). Commonw. Mycol. Inst. Mycol. Pap. 73:1-115.
Booth, C. 1961. Studies of Pyrenomycetes: VI. *Thielavia,* with notes on some allied genera. Commonw. Mycol. Inst. Mycol. Pap. 83:1-15.

Booth, C. 1971. The genus *Fusarium.* Commonw. Mycol. Inst., Kew. 237 pp.

Boyd, E. S. 1934. A developmental study of a new species of *Ophiodothella.* Mycologia 26:456-468.

Braun, U. 1987. A monograph of the Erysiphales (powdery mildews). J. Cramer, Berlin. 663 pp.

Cailleux, R. 1971. Recherches sur la mycoflore coprophile Centrafricaine. Les genres *Sordaria, Gelasinospora, Bombardia* (Biologie-Morphologie-Systématique); Écologie. Bull. Soc. Mycol. Fr. 87:461-626.

Cain, R. F. 1934. Studies of coprophilous Sphaeriales in Ontario. Univ. Toronto Stud. Biol. Ser. 38:1-126.

Cain, R. F. 1950. Studies of coprophilous Ascomycetes I. *Gelasinospora.* Can. J. Res., C 28:566-576.

Cannon, P. F. 1986. A revision of *Achaetomium, Achaetomiella* and *Subramaniula,* and some similar species of *Chaetomium.* Trans. Brit. Mycol. Soc. 87:45-76.

Cannon, P. F., and D. L. Hawksworth. 1982. A re-evaluation of *Melanospora* Corda and similar pyrenomycetes, with a revision of the British species. Bot. J. Linnean Soc. 84: 115-160.

Cannon, P. F., and D. L. Hawksworth. 1984. A revision of the genus *Neocosmospora* (Hypocreales). Trans. Brit. Mycol. Soc. 82: 673-688.

Cannon, P. F., and D. W. Minter. 1986. The Rhytismataceae of the Indian subcontinent. Commonw. Mycol. Inst. Mycol. Pap. 155:1-123.

Chardon, C. E., and R. A. Toro. 1930. Mycological explorations of Colombia. J. Dept. Agric. P. Rico 14: 195-369.

CMI. Descriptions of pathogenic fungi and bacteria. Commonwealth Mycol. Inst., Kew. (Since 1986 merged with IM and published first in Mycopathologia; also available as separate sets).

Corbaz, R. 1956. Recherches sur le genre *Didymella* Sacc. Phytopath. Z. 28:375-414.

Corlett, M. 1981. A taxonomic survey of some species of *Didymella* and *Didymella*-like species. Can. J. Bot. 59:2016-2042.

Czabator, F. J. 1976. A new species of *Ploioderma* associated with a pine needle blight. Mem. N. Y. Bot. Gard. 28:41-44.

Dargan, J. S., and K. S. Thind. 1979. Xylariaceae of India-VII. The genus *Rosellinia* in the Northeast Himalayas. Mycologia 71:1010-1023.

Darker, G. D. 1932. The Hypodermataceae of conifers. Contrib. Arnold Arb. 1:1-131.

Darker, G. D. 1967. A revision of the genera of the Hypodermataceae. Can. J. Bot. 45:1399-1444.

Denison, W. C. 1972. Central American Pezizales. IV. The genera *Sarcoscypha, Pithya,* and *Nanoscypha.* Mycologia 64:609-623.

Dennis, R. W. G. 1956. Some Xylarias of tropical America. Kew Bull. 3:401-444.

Dennis, R. W. G. 1958. Some Xylosphaeras of tropical Africa. Rev. Biol. 1:175-208.

Dennis, R. W. G. 1963. Remarks on the genus *Hymenoscyphus* S. F. Gray, with observations on sundry species referred by Saccardo and others to the genera *Helotium, Pezizella,* or *Phialea.* Persoonia 3:29-80.

Dennis, R. W. G. 1981. British Ascomycetes. Rev. ed. J. Cramer, Vaduz. 585 pp.

Diehl, W. W. 1950. *Balansia* and the Balansiae in America. U. S. Dept. Agric., Agric. Monogr. 4:1-82.

Dingley, J. M. 1951. The Hypocreales of New Zealand II. The genus *Nectria.* Trans. R. Soc. N. Z. Bot. 79:177-202.

Dingley, J. M. 1952a. The Hypocreales of New Zealand III. The genus *Hypocrea.* Trans. R. Soc. N. Z. 79:323-337.

Dingley, J. M. 1952b. The Hypocreales of New Zealand IV: The genera *Calonectria, Gibberella* and *Thyronectria.* Trans. R. Soc. N. Z. 79:403-411.

Doguet, G. 1955. Le genre *Melanospora:* biologie, morphologie, développement, systématique. Botaniste 39:1-313.

Doguet, G. 1960. Morphologie, organogénie et évolution nucléaire de l'*Epichloë typhina.* Bull. Soc. Mycol. Fr.

76:171-203.
Doi, Y. 1966. A revision of Hypocreales with cultural observation I. Some Japanese species of *Hypocrea* and *Podostroma.* Bull. Nat. Sci. Mus. Tokyo 9:345-357.
Doi, Y. 1972. Revision of the Hypocreales with cultural observations IV. The genus *Hypocrea* and its allies in Japan (2) Enumeration of the species. Bull. Nat. Sci. Mus. Tokyo 15:649-751.
Domsch, K. H., W. Gams, and T-H. Anderson. 1980. Compendium of soil fungi. Vol. 1. Academic Press, New York. 859 pp.
Ellis, J. B., and B. M. Everhart. 1892. The North American Pyrenomycetes. Publ. by the authors. Newfield, New Jersey. 793 pp. + 41 plates.
Ellis, M. B., and J. P. Ellis. 1985. Microfungi on Land Plants. Macmillan Publ. Co., New York. 818 pp.
Eriksson, O. 1967a. On graminicolous pyrenomycetes from Fennoscandia 1. Dictyosporous species. Ark. Bot. 6:339-380.
Eriksson, O. 1967b. On graminicolous pyrenomycetes from Fennoscandia 2. Phragmosporous and scolecosporous species. Ark. Bot. 6:381-440.
FC. Fungi Canadensis. Publ. by Biosystematics Research Inst., Agriculture Canada, Ottawa.
Francis, S. M. 1985. *Rosellinia necatrix* - fact or fiction? Sydowia 38:75-86.
Furuya, K., and S. Udagawa. 1972. Coprophilous Pyrenomycetes from Japan I. J. Gen. Appl. Microbiol. 18:433-454.
Furuya, K., and S. Udagawa. 1977. Coprophilous Pyrenomycetes from Japan IV. Trans. Mycol. Soc. Japan 17:248-261.
Gilkey, H. M. 1939. Tuberales of North America. Oregon St. Monogr., Stud. Bot. 1:1-63.
Gilkey, H. M. 1954. Tuberales. No. Amer. Flora, Ser. II(Pt. 1):1-36.
Glawe, D. A., and J. D. Rogers. 1984. Diatrypaceae in the Pacific Northwest. Mycotaxon 20:401-460.

Graham, J. H., and E. S. Luttrell. 1961. Species of *Leptosphaerulina* on forage plants. Phytopathology 51:680-693.

Griffin, H. D. 1966. The genus *Ceratocystis* in Ontario. Can. J. Bot. 46:689-718.

Groves, J. W., and S. C. Hoare. 1953. The Helvellaceae of the Ottawa District. Can. Field-Natl. 67:95-102.

Hanlin, R. T., and O. Tortolero. 1984. An unusual tropical powdery mildew. Mycologia 76:439-442.

Hanlin, R. T., and O. Tortolero. 1987. A new species and a new combination in *Cercophora.* Mycotaxon 30:407-416.

Hansford, C. G. 1961. The Meliolineae. A monograph. Sydowia Suppl. 2:1-806.

Hansford, C. G. 1963. Iconographia Meliolinearum. Sydowia Suppl. 5: 285 plates.

Harrington, T. C. 1981. Cyclohexamide sensitivity as a taxonomic character in *Ceratocystis*. Mycologia 73:1123-1129.

Hawksworth, D. L. 1971. A revision of the genus *Ascotricha* Berk. Commonw. Mycol. Inst. Mycol. Pap. 126:1-28.

Hawksworth, D. L., and S. Udagawa. 1977. Contributions to a monograph of *Microthecium.* Trans. Mycol. Soc. Japan 18:143-154.

Hodges, C. S., Jr. 1985. Hawaiian forest fungi VI. A new species of *Brasiliomyces* on *Sapindus oahuensis.* Mycologia 77:977-981.

Hodges, C. S., Jr., A. C. Alfenas, and F. A. Ferreira. 1986. The conspecificity of *Cryphonectria cubensis* and *Endothia eugeniae.* Mycologia 78:343-350.

Holm, L. 1952. Taxonomical notes on ascomycetes. II. The herbicolous Swedish species of the genus *Leptosphaeria* Ces. & De Not. Sven. Bot. Tidskr. 46:18-46.

Holm, L. 1953. Taxonomical notes on Ascomycetes. III. The herbicolous Swedish species of the genus *Didymella* Sacc. Sven. Bot. Tidskr. 47:520-525.

Holm, L. 1957. Études taxonomiques sur les Pléosporacées. Symb. Bot. Ups. 14(3):1-188.

Homma, Y. 1937. Erysiphaceae of Japan. J. Fac. Agric. Hokkaido Imp. Univ. 38:186-461.

Hoog, G. S. de, and R. J. Scheffer. 1984. *Ceratocystis* versus *Ophiostoma:* a reappraisal. Mycologia 76:292-299.

Hughes, S. J. 1976. Sooty moulds. Mycologia 48:693-820.

Hunt, J. 1956. Taxonomy of the genus *Ceratocystis.* Lloydia 19:1-58.

Hunt, R. S., and W. G. Ziller. 1978. Host-genus keys to the Hypodermataceae of conifer leaves. Mycotaxon 6:481-496.

IM. Iconographia Mycologica. Publ. as supplement to Mycopathologia et Mycologia Applicata. Since 1986 renamed Mycopathologia and IM merged with CMI descriptions.

Irwin, J. A. G., and R. D. Davis. 1985. Taxonomy of some *Leotosphaerulina* spp. on legumes in Eastern Australia. Aust. J. Bot. 33:233-237.

Jeng, R. S., E. R. Luck-Allen, and R. F. Cain. 1977. New species and new records of *Delitschia* from Venezuela. Can. J. Bot. 55:383-392.

Junell, L. 1967. Erysiphaceae of Sweden. Symb. Bot. Upsaliensis 19(1):1-117.

Kanouse, B. B. 1940. The genus *Plectania* and its segregates in North America. Mycologia 40:482-497.

Kempton, P. E., and V. L. Wells. 1970. Studies on the fleshy fungi of Alaska. IV. A preliminary account of the genus *Helvella.* Mycologia 62: 940-959.

Kobayashi, T. 1970. Taxonomic studies of Japanese Diaporthaceae with special reference to their life-histories. Bull. Govt. For. Exp. Sta. 226:1-242.

Kobayasi, Y., and D. Shimizu. 1983. [Iconography of Vegetable Wasps and Plant Worms.] Hoikuska Publ. Co., Ltd., Osaka. 280 pp. (in Japanese).

Kohn, L. M. 1979. A monographic revision of the genus *Sclerotinia.* Mycotaxon 9:365-444.

Kojwang, H. O., and T. Kurkela. 1984. *Linospora ceuthospora* on aspen (*Populus tremula*) in Finland. Karstenia 24:33-40.

Koponen, H., and K. Makela. 1975. *Leptosphaeria* s. lat. (*Keissleriella, Paraphaeosphaeria, Phaeosphaeria*) on Gramineae in Finland. Ann. Bot. Fenn. 12:141-160.

Krug, J. C., and R. F. Cain. 1974. A preliminary treatment of the genus *Podosordaria.* Can. J. Bot. 52:589-605.

Langdon, R. F. N. 1954. The origin and differentiation of *Claviceps* species. Univ. Queensland Pap., Dept. Bot. 3:61-68.

LeGal, M. 1958. Discomycètes du Maroc. I. Un *Urnula* nouveau: *Urnula megalocrater* Malençon et LeGal sp. nov. Etude de l'espèce, suivie d'une révision des caractères des genres *Urnula* et *Sarcosoma* Casp. Bull. Soc. Mycol. Fr. 74:155-177.

Leonard, K. J., and E. G. Suggs. 1974. *Setosphaeria prolata,* the ascigerous state of *Exserohilum prolatum.* Mycologia 66:281- 297.

Luck-Allen, E. R., and R. F. Cain. 1975. Additions to the genus *Delitschia.* Can. J. Bot. 53:1827-1887.

Lundqvist, N. 1972. Nordic Sordariaceae s. lat. Symbolae Bot. Ups. 20(1):1-374.

Luttrell, E. S., and C. W. Bacon. 1977. Classification of *Myriogenospora* in the Clavicipitaceae. Can. J. Bot. 55:2090- 2097.

Maas Geesteranus, R. A. 1972. *Spathularia* and *Spathulariopsis.* Proc. Kon. Ned. Akad. Wetensch., Ser. C 75:243-255.

Mains, E. B. 1954. North American species of *Geoglossum* and *Trichoglossum.* Mycologia 46:586-631.

Mains, E. B. 1955. North American hyaline-spored species of the Geoglosseae. Mycologia 47:846-877.

Mains, E. B. 1956. North American species of the Geoglossaceae. Tribe Cudonieae. Mycologia 48:694-710.

Mains, E. B. 1957. Species of *Cordyceps* parasitic on *Elaphomyces.* Bull. Torrey Bot. Club 84:243-251.

Mains, E. B. 1958. North American entomogenous species of *Cordyceps.* Mycologia 50:169-222.

Martin, P. 1967. Studies in the Xylariaceae: II. *Rosellinia* and the Primocinerea section of *Hypoxylon.* J. S. Afr. Bot. 33:315-328.

Martin, P. 1968a. Studies in the Xylariaceae: III. South African and foreign species of *Hypoxylon* sect. Entoleuca. J. S. Afr. Bot. 34:153-199.

Martin, P. 1968b. Studies in the Xylariaceae: IV. *Hypoxylon,* sections Papillata and Annulata. J. S. Afr. Bot. 34:303-330.
Martin, P. 1968c. Studies in the Xylariaceae: V. *Euhypoxylon.* J. S. Afr. Bot. 35:149-206.
Martin, P. 1969. Studies in the Xylariaceae: VI. *Daldinia, Nummulariola* and their allies. J. S. Afr. Bot. 35:267-320.
Martin, P. 1970. Studies in the Xylariaceae: VIII. *Xylaria* and its allies. J. S. Afr. Bot. 36:73-138.
Matsushima, T. 1971. Microfungi of the Solomon Islands and Papua-New Guinea. Pub. by author, Kobe. 84 pp. + 169 figures and 48 plates.
Matsushima, T. 1975. Icones Microfungorum a Matsushima Lectorum. Pub. by author, Kobe. 209 pp. + 415 plates.
Micales, J. A., and R. J. Stipes. 1987. A reexamination of the fungal genera *Cryphonectria* and *Endothia.* Phytopathology 77:650-654.
Miller, J. H. 1940. The genus *Myriangium* in North America. Mycologia 32:587-600.
Miller, J. H. 1961. A Monograph of the World Species of *Hypoxylon.* Univ. Georgia Press, Athens. 158 pp.
Miller, J. H., and F. A. Wolf. 1936. A leaf-spot disease of honey locust caused by a new species of *Linospora.* Mycologia 28:171-180.
Minter, D. W. 1981. *Lophodermium* on pines. Commonw. Mycol. Inst. Mycol. Pap. 147:1-54.
Minter, D. W., and M. P. Sharma. 1982. Three species of *Lophodermium* from the Himalayas. Mycologia 74:702-711.
Monod, M. 1983. Monographie taxonomique des Gnomoniaceae. Beihefte Sydowia, Ser. II. 9:1-315.
Moreau, C. 1953. Les genres *Sordaria* et *Pleurage.* Leurs affinités systématiques. Encyc. Mycol. 25:1-330.
Morton, F. J., and G. Smith. 1963. The genera *Scopulariopsis* Bainier, *Microascus* Zukal, and *Doratomyces* Corda. Commonw. Mycol. Inst. Mycol. Pap. 86:1-96.
Müller, E. 1950. Die schweizerischen Arten der Gattung *Leptosphaeria* und ihrer Verwandten. Sydowia 4:185-319.
Müller, E. 1951. Die schweizerischen Arten der Gattungen *Clathrospora, Pleospora, Pseudoplea* und *Pyrenophora.*

Sydowia 5:248- 310.
Müller, E., and J. A. von Arx. 1962. Die Gattungen der didymosporen Pyrenomyceten. Beitr. Kryptogamenfl. Schweiz 11(2):1-922.
Munk, A. 1957. Danish Pyrenomycetes. A Preliminary Flora. Dansk Bot. Arkiv 17(1):1-491.
Olchowecki, A., and J. Reid. 1974. Taxonomy of the genus *Ceratocystis* in Manitoba. Can. J. Bot. 52:1675-1711.
Orton, C. R. 1944. Graminicolous species of *Phyllachora* in North America. Mycologia 36:18-53.
Parbery, D. G. 1967. Studies on graminicolous species of *Phyllachora* Nke. in Fckl. V. A taxonomic monograph. Aust. J. Bot. 15:271-375.
Parker, A. K. 1957. *Europhium,* a new genus of the Ascomycetes with a *Leptographium* imperfect state. Can. J. Bot. 35:173-179.
Parmalee, J. A. 1977. The fungi of Ontario. II. Erysiphaceae (mildews). Can. J. Bot. 55:1940-1983.
Pérez-Silva, E. 1973. El género *Daldinia* (Pyrenomycetes) en México. Bol. Soc. Mex. Mic. 7:51-58.
Pérez-Silva, E. 1975. El género *Xylaria* (Pyrenomycetes) en México, I. Bol. Soc. Mex. Mic. 9:31-52.
Petrini, L., and O. Petrini. 1985. Xylariaceous fungi as endophytes. Sydowia 38:216-234.
Petrini, L. E., and J. D. Rogers. 1986. A summary of the *Hypoxylon serpens* complex. Mycotaxon 26:401-436.
Poroca, D. J. M. 1986. Revisão histórica das Xylariaceae do Brasil. Bol. Micol. 3:41-53.
Punithalingam, E. 1974. Studies on Sphaeropsidales in culture. II. Commonw. Mycol. Inst., Mycol. Pap. 136:1-63.
Rappaz, F. 1987. Taxonomy and nomenclature of the octosporous Diatrypaceae. Mycol. Helvet. 2:285-648.
Rifai, M. A. 1968. The Australasian Pezizales in the herbarium of The Royal Botanic Gardens Kew. Verhdl. Koninkl. Nederl. Akad. Wetensch. Naturk. 57(3):1-295.
Roane, M. K., G. J. Griffin, and J. R. Elkins. 1986. Chestnut blight, other Endothia diseases, and the genus *Endothia.* APS Press, St. Paul. 53 pp.
Roberts, R. G., J. A. Robertson, and R. T. Hanlin. 1984.

Ascotricha xylina: its occurrence, morphology, and typification. Mycologia 76:963-968.
Rogers, J. D. 1986. Provisional keys to *Xylaria* species in continental United States. Mycotaxon 26:85-97.
Rogers, J. D., and B. E. Callan. 1986. *Xylaria polymorpha* and its allies in continental United States. Mycologia 78:391-400.
Rogerson, C. T., and G. J. Samuels. 1985. Species of *Hypomyces* and *Nectria* occurring on Discomycetes. Mycologia 77:763-783.
Rossman, A. Y. 1979a. *Calonectria* and its type species, *C. daldiniana,* a later synonym of *C. pyrochroa.* Mycotaxon 8:321-328.
Rossman, A. Y. 1979b. A preliminary account of the taxa described in *Calonectria.* Mycotaxon 8:485-558.
Rossman, A. Y. 1983. The phragmosporous species of *Nectria* and related genera. Commonw. Mycol. Inst. Mycol. Pap. 150:1-164.
Saccas, A. M. 1956. Les *Rosellinia* des caféiers en Oubangui-Chari. Agron. Trop. 11:551-614.
Samuels, G. J. 1976. A revision of the fungi formerly classified as *Nectria* subgenus *Hyphonectria.* Mem. N. Y. Bot. Gard. 26(3): 1-126.
Samuels, G. J., and K. P. Dumont. 1982. The genus *Nectria* (Hypocreaceae) in Panama. Caldasia 13:372-423.
Sandu-Ville, C. 1967. Ciupercile Erysiphaceae din România. Ed. Acad. Rep. Soc. România, Bucarest. 358 pp.
Scheinpflug, H. 1958. Untersuchungen über die Gattung *Didymosphaeria* Fuck. und einige verwandte Gattungen. Ber. Schweiz Bot. Ges. 68:325-385.
Schuepp, H. 1959. Untersuchungen über *Pseudopezizoideae* sensu Nannfeldt. Phytopath. Z. 36:213-269.
Schumacher, T. 1978. A guide to the amenticolous species of the genus *Ciboria* in Norway. Norw. J. Bot. 25:145-155.
Seaver, F. J. 1910. The Hypocreales of North America-III. Mycologia 2:48-92.
Seaver, F. J. 1942. North American Cup-Fungi (Operculates). Pub. by author, New York. 377 pp.
Seaver, F. J. 1951. The North American Cup-fungi

(Inoperculates). Pub. by author, New York. 428 pp.
Seaver, F. J., and C. E. Chardon. 1926. Botany of Porto Rico and the Virgin Islands - Mycology. N. Y. Acad. Sci. Scient. Surv. P. Rico & Virgin Isls. 8(1):1-208.
Seth, H. K. (1970)1972. A monograph of the genus *Chaetomium.* Nova Hedw. Beih. 37:1- 133.
Shoemaker, R. A. 1984. Canadian and some extralimital *Leptosphaeria* species. Can. J. Bot. 62:2688-2729.
Simmons, E. G. 1986. *Alternaria* themes and variations (22-26). Mycotaxon 25: 287-308.
Sivanesan, A. 1977. The taxonomy and pathology of *Venturia* species. J. Cramer, Vaduz. 139 pp.
Sivanesan, A. 1984. The bitunicate ascomycetes and their anamorphs. J. Cramer, Vaduz. 701 pp.
Skolko, A. J., and J. W. Groves. 1948. Notes on seed-borne fungi V. *Chaetomium* species with dichotomously branched hairs. Can. J. Res. C 26:269-280.
Skolko, A. J., and J. W. Groves. 1953. Notes on seed-borne fungi VII. *Chaetomium.* Can. J. Bot. 31:779-809.
Speer, E. O. 1973. Untersuchungen zur Morphologie und Systematik der Erysiphaceen I. Die Gattung *Blumeria* Golovin und ihre Typusart *Erysiphe graminis* DC. Sydowia 27:1-6.
Stevens, F. L. 1924. Parasitic fungi from British Guiana and Trinidad. Illinois Biol. Monogr. 8(3):1-76.
Stevens, F. L. 1927. Fungi from Costa Rica and Panama. Illinois Biol. Monogr. 11(2):1-103.
Swart, H. J. 1982. Australian leaf-inhabiting fungi XV. *Ophiodothella longispora* sp. nov. Trans. Brit. Mycol. Soc. 79:566-568.
Tehon, L. R. 1935. A monographic rearrangement of *Lophodermium.* Illinois Biol. Monogr. 13:1-151.
Thind, K. S., and J. S. Dargan. 1978. Xylariaceae of India - IV The genus *Daldinia.* Kavaka 6:15-24.
Tubaki, K. 1960. An undescribed species of *Hypomyces* and its conidial stage. Nagaoa 7:29-34.
Tubaki, K. 1966. An undescribed species of *Hymenoscyphus,* a perfect stage of *Varicosporium.* Trans. Brit. Mycol. Soc. 49:345-349.

Tubaki, K. 1975. *Hypomyces* and its conidial states in Japan. Rept. Tottori Mycol. Inst. 12:161-169.

Udagawa, S. 1960. A taxonomic study on the Japanese species of *Chaetomium.* J. Gen. Appl. Microbiol. 6:223-251.

Upadhyay, H. P. 1981. A monograph of *Ceratocystis* and *Ceratocystiopsis.* Univ. Georgia Press, Athens. 176 pp.

Viégas, A. P. 1944. Alguns fongos do Brasil II. Ascomicetos. Bragantia 4:1-392.

Ueda, S., and S. Udagawa. 1983. A new Japanese species of *Neocosmospora* from marine sludges. Mycotaxon 16:387-395.

Walker, J. C. 1972. Type studies on *Gaeumannomyces graminis* and related fungi. Trans. Brit. Mycol. Soc. 58:427-457.

Walker, J. 1980. *Gaeumannomyces, Linocarpon, Ophiobolus* and several other genera of scolecospored ascomycetes and *Phialophora* conidial states, with a note on hyphopodia. Mycotaxon 11:1-129.

Weber, N. S. 1972. The genus *Helvella* in Michigan. Mich. Bot. 11:147-210.

Wehmeyer, L. E. 1933. The genus *Diaporthe* Nitschke and its segregates. Univ. Michigan Press, Ann Arbor. 349 pp.

Wehmeyer, L. E. 1961. A world monograph of the Genus *Pleospora* and its segregates. Univ. Michigan Press, Ann Arbor. 451 pp.

Weijman, A. C. M., and G. S. de Hoog. 1975. On the subdivision of the genus *Ceratocystis.* Ant. Leeuwenhoek 41:353-360.

Whalley, A. J. S., and G. N. Greenhalgh. 1973. Numerical taxonomy of *Hypoxylon.* II. A key for the identification of British species of *Hypoxylon.* Trans. Brit. Mycol. Soc. 61:455-459.

Whetzel, H. H., and N. F. Buchwald. 1936. North American species of *Sclerotinia* and related genera. III. *Ciboria acerina.* Mycologia 28:514-527.

Whetzel, H. H., and F. A. Wolf. 1945. The cup fungus, *Ciboria carunculoides,* pathogenic on mulberry fruits. Mycologia 37:476-491.

Williamson, M. A., and E. C. Bernard. 1988. Life cycle of a new species of *Blumeriella* (Ascomycotina: Dermateaceae), a leaf-spot pathogen of spiraea. Can. J. Bot. 66:2048-2054.

Zheng, R-y. 1981. The genus *Erysiphe* in China. Sydowia 34:214-327.

Zheng, R-y. 1984. The genus *Brasiliomyces* (Erysiphaceae). Mycotaxon 19:281-289.

Zheng, R-y., and G-q. Chen. 1977a. [Taxonomic studies on the genus *Uncinula* of China I. Discussion on *Uncinula sinensis* Tai et Wei.] Acta Microbiol. Sinica 17:189-197. (in Chinese).

Zheng, R-y., and G-q. Chen. 1977b. [Taxonomic studies on the genus *Uncinula* of China III. New species and new varieties on Coriariaceae, Euphorbiaceae, Oleaceae, Sterculiaceae and Tiliaceae.] Acta Microbiol. Sinica 17: 281-292. (in Chinese).

Zheng, R-y., and G-q. Chen. 1978. [Taxonomic studies on the genus *Uncinula* of China IV. New species, new variety and new combination on Hamameliadaceae, Lauraceae, Moraceae and Ulmaceae.] Acta Microbiol. Sinica 18: 11-22. (in Chinese).

Zheng, R., and G. Chen. 1981. [Taxonomic studies on the genus *Uncinula* of China V. New species and new variety on Hamameliadaceae, Papillionaceae, Rutaceae and Salicaceae.] Acta Microbiol. Sinica 21: 298-307. (in Chinese).

Zogg, H. 1962. Die Hysteriaceae s. str. und Lophiaceae. Beitr. Kryptogamenfl. Schweiz 11(3):1-190.

INDEX TO GENERA AND SPECIES